Second Report to Congress:

Highlights of the Diesel Emissions Reduction Program

Prepared by:

U.S. EPA's Office of Transportation and Air Quality

Washington, DC

Contents

Executive Summary

From goods movement to building construction to public transportation, diesel engines are the modern-day workhorse of the American economy. Diesel engines are extremely efficient, and they power nearly every major piece of machinery and equipment on farms, on construction sites, in ports, and on highways. However, not all diesel engines are as clean as those manufactured after 2006 and later, when EPA's stringent heavy-duty highway and non-road engine standards began coming into effect. EPA estimates that approximately 11 million older diesel engines remain in use, and will continue to emit significant amounts of nitrogen oxides (NO_x) and particulate matter (PM) until they wear out and are replaced. To reduce the public's exposure to pollution from these older, dirtier engines, Congress in 2005 authorized funding for the Diesel Emissions Reduction Act, a grant program designed to selectively retrofit or replace the older diesel engines most likely to impact human health. The U.S. Environmental Protection Agency (EPA) administers all Diesel Emissions Reduction Act (DERA) funding under the umbrella of the National Clean Diesel Campaign (NCDC), which promotes clean air strategies by working with manufacturers, fleet operators, air quality professionals, environmental and community organizations, and state and local officials to reduce diesel emissions.

For more information about the background of NCDC, please see the first *Report to Congress: Highlights of the Diesel Emissions Reduction Program*, EPA-420-R-09-006.

DERA Funding Has Provided a Broad Range of Benefits

Since 2008, the DERA program has achieved impressive outcomes and a range of benefits, such as:

➲ **Improved air quality, health benefits, and fuel savings.** EPA grants have funded projects that provided immediate health and environmental benefits. From 2008 to 2010, EPA awarded nearly $470 million to more than 350 grantees in 50 states and the District of Columbia to retrofit, replace, or repower more than 50,000 vehicles and equipment in a variety of industries. EPA estimates that these projects will reduce emissions by at least 203,900 tons of NO_x and 12,500 tons of PM over the lifetime of the affected engines. As a result of these pollution reductions, EPA estimates that the health benefits associated with up to 1,400 fewer premature deaths and fewer hospital visits, among other impacts, will total approximately $3.4 billion to $8.2 billion.[1] These clean diesel projects also are estimated to reduce carbon monoxide (CO) emissions by 48,000 tons, hydrocarbon (HC) emissions by 18,000 tons, and carbon dioxide (CO_2) emissions by 2.3 million tons, as well as generate fuel savings of over 205 million gallons as a result of idle reduction.

➲ **Cleaned up the nation's supply chain.** Along with EPA's SmartWay Transport Partnership program, DERA funding has focused on diesel pollution at intermodal hubs, such as delivery centers and ports, and across the nation's transportation infrastructure that supplies goods. More than $300 million in funding has been targeted to reducing emissions from the nation's supply chain. NCDC will work closely with the SmartWay Program as part of the Legacy Fleet Plan in the future to target areas with high diesel emissions.

---

[1] EPA's estimates for health benefits assume that each avoided premature death is worth the value of statistical life (VSL). EPA recommends use of the central estimate for VSL of $7.4 million ($2006), updated to the year of the analysis, be used in all benefits analyses that seek to quantify mortality risk reduction benefits regardless of the age, income, or other population characteristics of the affected population.

➲ **Generated economic and environmental activity.** Clean diesel projects are cost-effective, according to EPA's calculations of health benefits. Each federal dollar invested in clean diesel projects has leveraged as much as $3 from other government agencies, private organizations, industry, and nonprofit organizations, generating between $7 and $18 in public health benefits. In addition, new clean diesel technologies can spur environmental jobs and innovation in the marketplace.

➲ **Answered popular demand.** Stakeholders have shown a tremendous amount of interest in EPA-funded clean diesel projects. Funding requests have exceeded availability by as much as 7:1.

➲ **Met local needs.** These grants have solved local problems with locally conceived solutions—all DERA grant recipients have tailored their projects to their specific community's needs.

➲ **Served environmental justice communities.** Many projects have made health and environmental impacts in socially vulnerable areas. Rail and port projects are especially beneficial because they tend to take place in environmental justice communities, which are disproportionately impacted by higher levels of diesel exhaust.

Fiscal Year 2008

In its inaugural year, EPA funded 119 projects with $49 million. These funds retrofitted over 14,000 vehicles, vessels, or pieces of equipment. This included almost 6,000 school buses and nearly 4,500 long-haul trucks. The Emerging Technologies (ET) program, an innovative DERA sub-program with a goal of testing the effectiveness of new clean diesel technologies in the field, got its start in 2008. Through this program, EPA verified its first emission control devices for marine vessels and locomotives. One such technology was Caterpillar's marine engine upgrade kit for certain engine models, which is now available for widespread use to help ship owners reduce pollution. EPA estimates that this first cohort of 119 projects will reduce 22,700 tons of NO_x, 2,700 tons of PM, 4,200 tons of HC, 15,900 tons of CO, and 289,900 tons of CO_2 over the lifetime of the affected engines. These pollution reductions, according to EPA estimates, will provide approximately $644 million to $1.6 billion in health benefits associated with fewer premature deaths and other health impacts, and save nearly 26 million gallons of fuel as a result of projects to reduce engine idling.

American Recovery and Reinvestment Act

EPA received a $300 million appropriation under the American Recovery and Reinvestment Act (Recovery Act) in 2009 and awarded funds to 160 clean diesel projects. Every state and the District of Columbia received funding, and an additional 109 competitive projects were awarded funds across the country. EPA's first Recovery Act grant went to the state of Colorado on March 27, 2009, for a school bus project that retrofitted nearly 1,000 buses. Important considerations in awarding Recovery Act grants were that the projects be "shovel-ready" and immediately create and retain American jobs. Through the Diesel Emissions Reduction Program, this funding resulted in the retention and creation of over 3,000 jobs.[2] In total, EPA estimates that the clean diesel Recovery Act projects will reduce 130,600 tons of NO_x, 7,200 tons of PM, 10,200 tons of HC, 22,800 tons of CO, and 1,309,100 tons of CO_2, as well as save over 116 million gallons of fuel over the lifetime of the affected engines. The health benefits of these pollution reductions equate to approximately $2 billion to $4.9 billion.

Fiscal Year 2009/2010

EPA received $60 million appropriations in both 2009 and 2010 and combined those funds to award 84 grants. One grant to Mississippi funded the retrofit or replacement of over 2,000 buses, and provided the leverage for Mississippi to raise additional funds for diesel retrofits and replacements, allowing the

---

[2] This jobs estimate was created based on self-reported information from Recovery Act grant recipients according to the Office of Management and Budget's guidance on job reporting.

state to reach its goal of retrofitting nearly every eligible school bus. Other applicants sought funding for large and impactful projects in under-served communities, such as those around ports and rail yards. Projects to repower or replace these heavy-emitting engines brought multi-pollutant reductions to areas disproportionately affected by diesel exhaust. Applicants also wanted to save fuel and lower emissions on long-haul trucks. EPA estimates that the fiscal year (FY) 2009/2010 grants will lead to emissions reductions of 50,600 tons of NO_x, 2,600 tons of PM, 3,600 tons of HC, 9,300 tons of CO, and 706,000 tons of CO_2 over the lifetime of the affected engines, with fuel savings of nearly 63 million gallons associated with projects to reduce idling. EPA estimates that these pollution reductions generate health benefits of $728 million to $1.8 billion.

Through these three award competitions, EPA has funded clean diesel projects from a variety of sectors to achieve emissions reductions across the country. In total, more than 50,000 pieces of equipment and vehicles have been retrofitted, replaced, repowered, or aided with idling reduction equipment so that they contribute fewer emissions to our air.

DERA Funding Has Targeted Areas of Greatest Need

Per the priorities outlined in its authorizing legislation, the DERA program has placed emphasis on maximizing health benefits and serving areas with poor air quality, such as areas of non-attainment for PM and ozone in its funding competitions. Approximately 70% of competitive projects have taken place in nonattainment areas for $PM_{2.5}$ (per the 2006 standard) or 8-hour ozone (per the 2008 standard).[3] Many projects in areas already in attainment of air quality standards targeted sensitive populations, such as children, by funding replacements and retrofits for school buses. Finally, in the last two years, the program has increased its work in ports. More than 70 projects in ports have been funded, addressing nearly 2,000 vehicles or equipment.

Fostered Clean Diesel Projects on Tribal Lands, Territories, and Islands

Tribal grants are another important component of the program's commitment to environmental justice. EPA has awarded more than $2 million for eight tribal grants in Minnesota, California, Alaska, Arizona, Washington, and Iowa. The tribal projects have retrofitted or replaced school buses, refuse haulers, transport buses, utility vehicles, fishing vessels, and mining equipment.

In FY 2009, DERA began funding projects in the District of Columbia and added Puerto Rico in FY 2011. DERA's most recent authorization, which took effect on October 1, 2011, allows Guam, the United States Virgin Islands, American Samoa, and the Commonwealth of the Northern Mariana Islands to receive federal funds for clean diesel projects.

Informed Communities

NCDC works to keep Americans informed of the benefits of clean diesel projects by increasing awareness. DERA-funded projects appear on EPA's website so that the public can learn more about clean diesel projects in their local areas. Project partners can install a *MyEnvironment* "widget" on their own websites so their users can learn more about projects in their communities. Additionally, NCDC expanded on past work with school bus retrofits and Scholastic by releasing a Spanish translation of *The Magic School Bus Gets Cleaned Up* (*El Autobús Mágico Necesita una Limpieza*).

---

[3] The percentage of projects taking place in non-attainment areas was calculated using EPA's Office of Air and Radiation's most recent National Ambient Air Quality Standards, which can be found at www.epa.gov/air/criteria.html.

FUNDING HAS SUPPORTED A WIDE VARIETY OF SECTORS AND TECHNOLOGIES

Retrofitted Equipment Across All Diesel Engine Sectors

Since its inception the program has focused on a wide variety of fleets (see Figure A, "Equipment Retrofitted by Fiscal Year and Sector"). Many projects have focused on school buses due to the importance of cleaner transportation for children and communities. In addition, the program targeted long-haul trucks, assisting fleets with fuel-saving technologies, and lowering emissions.

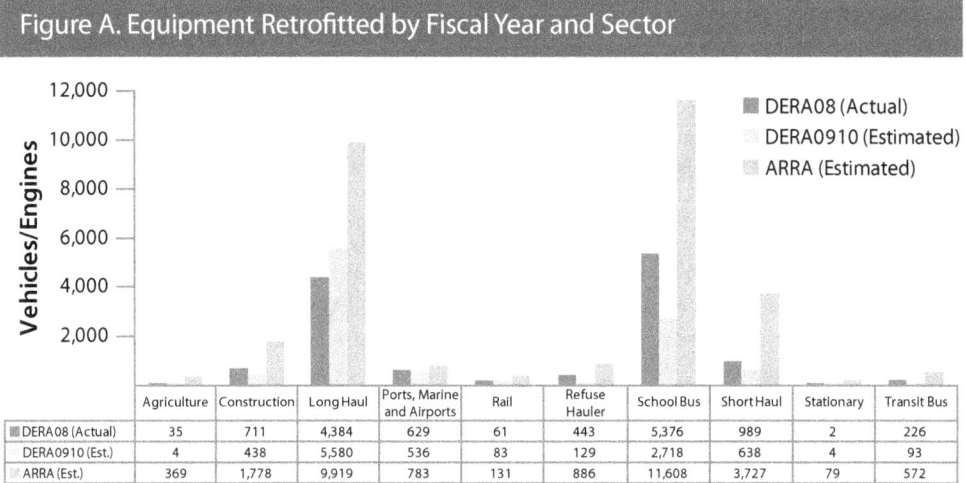

Figure A. Equipment Retrofitted by Fiscal Year and Sector

	Agriculture	Construction	Long Haul	Ports, Marine and Airports	Rail	Refuse Hauler	School Bus	Short Haul	Stationary	Transit Bus
DERA08 (Actual)	35	711	4,384	629	61	443	5,376	989	2	226
DERA0910 (Est.)	4	438	5,580	536	83	129	2,718	638	4	93
ARRA (Est.)	369	1,778	9,919	783	131	886	11,608	3,727	79	572

EPA Has Matched Technologies and Engines to Achieve Emissions Reductions

EPA has employed a variety of technologies to reduce emissions. One strategy is to retrofit vehicles with aftermarket technologies, such as diesel oxidation catalysts, closed crankcase ventilation, diesel particulate filters, or selective catalytic reduction. Other key strategies include installing idling reduction technologies such as auxiliary power units and direct fired heaters, tires and trailer skirts, and shore power or truck-stop electrification, and using cleaner fuels. Replacing or repowering older engines or vehicles is another important way to reduce emissions. The numerous technology options allow the DERA program to award funds to recipients based on the most cost-effective application of the technology. See Figure B, "Equipment by Fiscal Year and Technology," for the range of technologies used on engines.

From FY 2008 to the Recovery Act, applications requesting grants for repower and replacement projects increased by 25 percent. DERA funding for replacement and repowers is especially helpful to small business owners since it makes vehicle upgrades and emissions reductions more affordable. Many of these replacements, especially in the marine, port, and rail sectors, take place in environmental justice areas and make across-the-board diesel emissions reductions—lowering NO_x, PM, HC, CO, and CO_2 pollution.

Innovative Technologies

The ET program has fostered the development of cutting-edge, next generation diesel emissions reduction technologies by partnering technology manufacturers with fleets to test the effectiveness of the products and, if appropriate, become verified, making them available for wider use. The program has supported projects to demonstrate and improve 17 different technologies since 2008. In the past

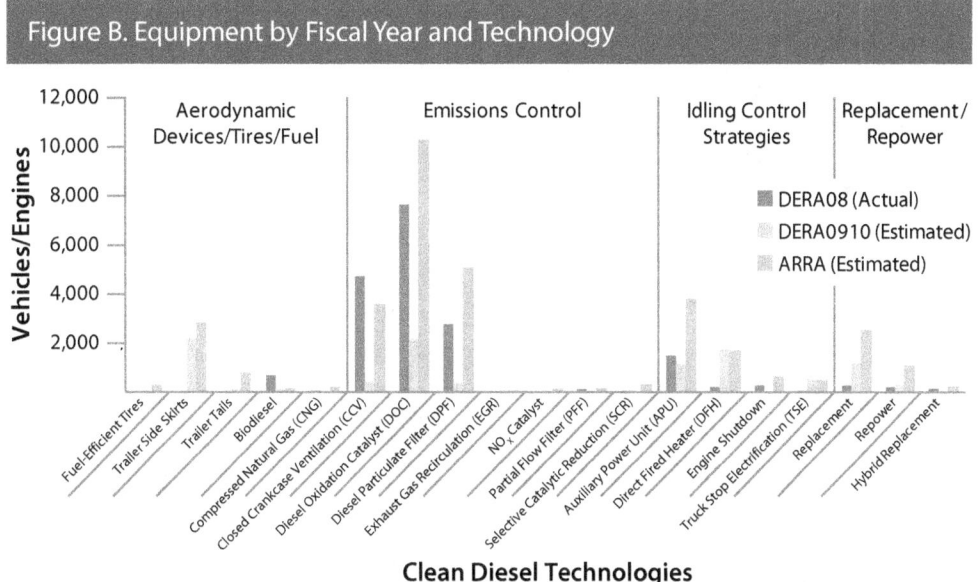

Figure B. Equipment by Fiscal Year and Technology

two years, two technologies previously on the emerging technologies list stand out as contributing to new technology options for fleets: a selective catalytic reduction (SCR) system for non-road equipment and an upgrade kit for marine engines.

LOOKING FORWARD

Funding for the DERA program has created environmental and public health benefits across the country. The most recent funds awarded in FY 2011 ($50 million) and in FY 2012 ($30 million) will continue to be used to reduce diesel emissions and deliver local environmental benefits. In December of 2010, Congress unanimously reauthorized the DERA program. On January 4, 2011, President Barack Obama signed DERA's reauthorization through 2016, allowing up to $100 million in annual appropriations. Over the course of the past three years, EPA has worked to improve efficiency and coordinate with project partners to manage unforeseen obstacles. EPA has managed product delivery delays, revised its Requests for Proposals, and refined grant tracking processes and databases. Additionally, EPA's Office of Inspector General has conducted assessments on the DERA program and offered recommendations to improve management and oversight to ensure that projects achieve the planned emissions reductions.

The DERA program has shown that retrofits and engine replacements are effective in reducing emissions and provided valuable lessons in how to administer clean diesel programs. Going forward, EPA plans to sharpen its focus on any remaining areas of disproportionate exposure to emissions from diesel engines, and ensure that clean diesel projects are as cost-effective as possible.

In FY 2013 the Agency will pilot a new approach that will target specific fleets in high diesel exposure areas such as near ports and freight distribution hubs and other disproportionately affected communities. The new strategy would allocate funds to a new rebate program established under DERA's reauthorization, and to grants, in part, for finance mechanisms that help fleets reduce diesel emissions. EPA believes the rebate, grant, and finance programs will allow greater precision in scrapping certain model years of vehicles and equipment and assisting public and private fleet owners with retrofitting or replacing those engines.

As the program matures, and subject to the availability of funds, EPA will build on past experience to ensure that clean diesel projects are as cost-effective as possible while targeting high exposure areas with the most effective technologies.

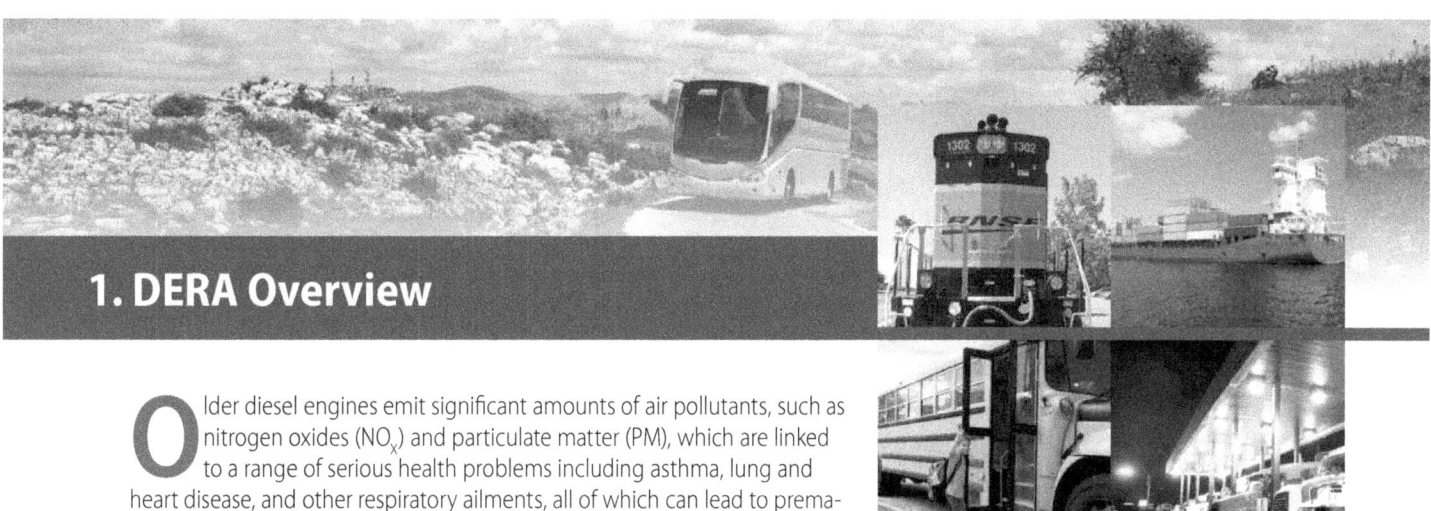

1. DERA Overview

Older diesel engines emit significant amounts of air pollutants, such as nitrogen oxides (NO_x) and particulate matter (PM), which are linked to a range of serious health problems including asthma, lung and heart disease, and other respiratory ailments, all of which can lead to premature death. Operating throughout our transportation infrastructure today, 11 million existing diesel engines—the nation's "legacy fleet"—will be retired over time, but some will remain in use for up to 20 or more years.

Figure 1. Diesel Engine Turnover by Year[4]

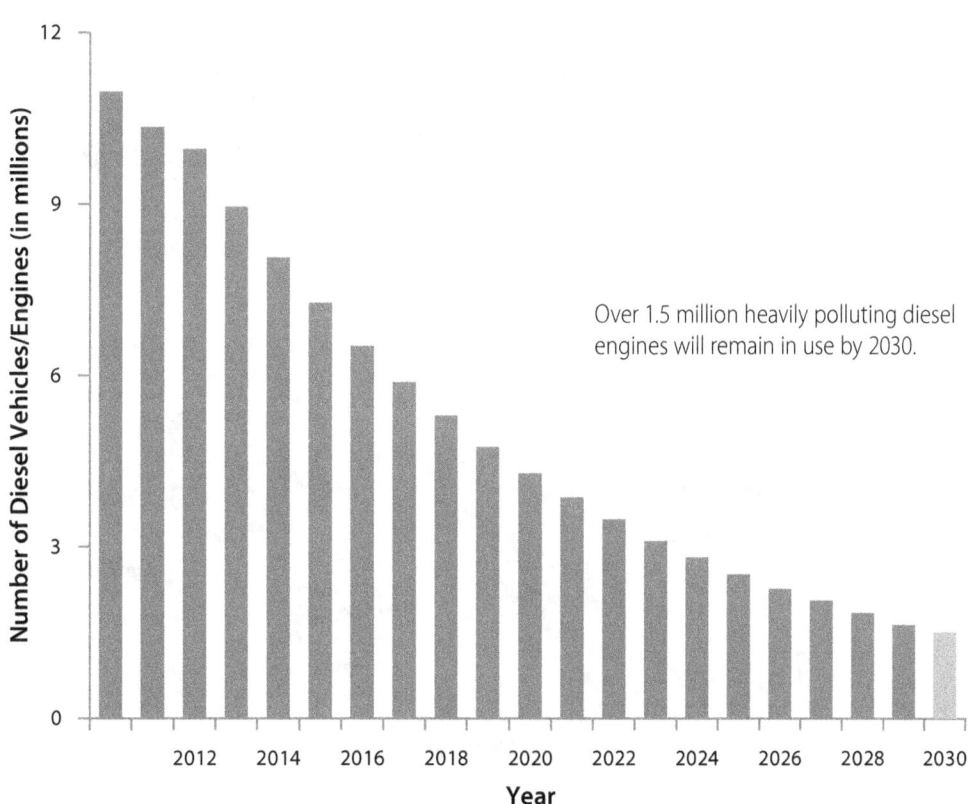

Over 1.5 million heavily polluting diesel engines will remain in use by 2030.

Y-axis: Number of Diesel Vehicles/Engines (in millions)

X-axis: Year — 2012, 2014, 2016, 2018, 2020, 2022, 2024, 2026, 2028, 2030

---

[4] This graph was created according to the MOVES and NONROAD models, which can be found at www.epa.gov/otaq/models/moves/movesback.htm and www.epa.gov/omswww/nonrdmdl.htm. Data based on a projected 10 percent fleet turnover rate from EPA modeling.

Recognizing that many states, such as California, Texas, Washington, Illinois, Maine, and New York, have developed successful clean diesel programs over the years, Congress authorized EPA's Diesel Emissions Reduction Act (DERA) program to provide funds to states directly, as well as to establish a national competitive grant program. Administered by EPA's National Clean Diesel Campaign (NCDC), the fiscal year (FY) 2008 competitive awards funded 119 projects. Congress then appropriated $300 million for the DERA program as part of the American Recovery and Reinvestment Act (Recovery Act). These appropriations funded an additional 160 projects. In FY 2009/2010, NCDC streamlined its award process and subsequently awarded 84 grants.

Taken together, EPA estimates that projects funded with FY 2008–2010 funds will reduce at least 203,900 tons of NO_x and 12,500 tons of PM over the lifetime of the affected engines, leading to estimated health benefits of between $3.4 billion and $8.2 billion. A significant part of the monetized value associated with these health benefits derives from EPA's estimates of the number of premature deaths avoided as a result of reduced exposure to PM and NO_x emissions.[5] As part of the health benefits calculation, EPA estimates that the PM and NO_x emissions reductions will lead to 540 to 1,400 prevented premature deaths. EPA also estimates that these clean diesel projects will reduce lifetime emissions of carbon monoxide (CO) by 48,000 tons, hydrocarbon (HC) by 18,000 tons, and carbon dioxide (CO_2) by 2,300,000 tons over the lifetime of the affected engines, as well as save over 205 million gallons of fuel as a result of projects that discourage idling and reliance on diesel engines to generate electricity.

Figure 2. Equipment Retrofitted, Replaced, or Repowered by DERA 2008 and 2009/2010 Grants

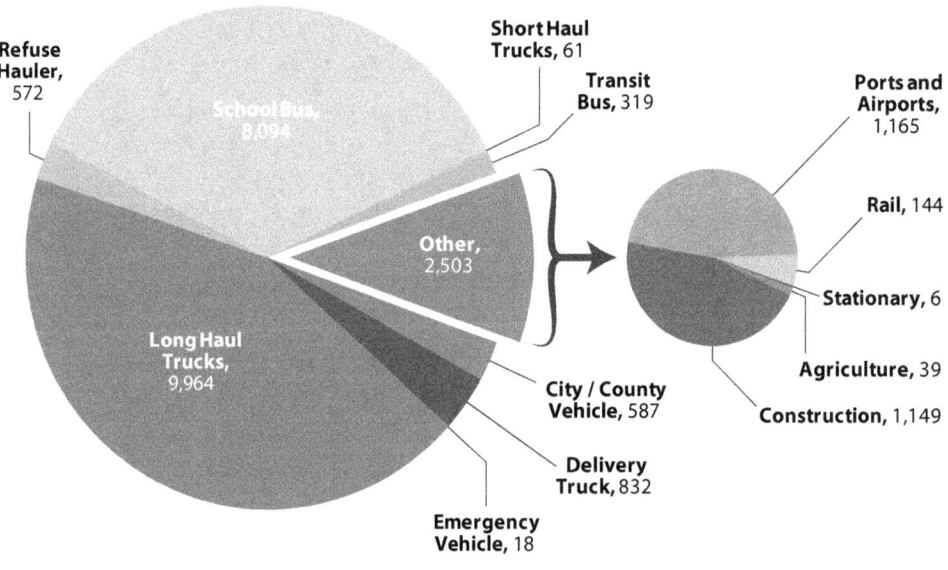

Refuse Hauler, 572

School Bus, 8,094

Short Haul Trucks, 61

Transit Bus, 319

Ports and Airports, 1,165

Rail, 144

Other, 2,503

Stationary, 6

Long Haul Trucks, 9,964

Agriculture, 39

City / County Vehicle, 587

Construction, 1,149

Delivery Truck, 832

Emergency Vehicle, 18

---

[5] EPA's estimates for health benefits assume that each avoided premature death is worth the value of a statistical life (VSL). EPA recommends use of the central estimate for VSL of $7.4 million ($2006), updated to the year of the analysis, be used in all benefits analyses that seek to quantify mortality risk reduction benefits regardless of the age, income, or other population characteristics of the affected population.

Grant Funding Has Delivered Multiple Benefits

Improved Air Quality and Generated Health Benefits

When grantees have retrofitted, rebuilt, or repowered engines or vehicles; switched to cleaner fuels; installed idle reduction technologies; or pursued another clean diesel strategy, the resulting air quality benefits have been immediate. For every dollar EPA has invested in clean diesel, the public has received $7 to $18 in health benefits, depending on the particular type of vehicle and technology combination. Additionally, grants have attracted matching funds from other government agencies, private organizations, industry, and nonprofit organizations, which have invested as much as $3 for every dollar provided by the grant.

For example, the North Carolina Department of Environment and Natural Resources was able to expand its $1.7 million award into a total investment of $9.3 million. This project repowered 14 marine vessels, installed 160 auxiliary power units, purchased 55 model year 2010 heavy duty trucks, retrofitted 186 school buses with diesel oxidation catalysts and closed crankcase ventilation systems, and funded 10 other equipment replacements and retrofits.

Assisted State and Local Governments

Funding has assisted state and local governments, which need to demonstrate compliance with federal air quality regulations such as the National Ambient Air Quality Standards for PM and ozone. These agencies have used funding to reduce diesel emissions, while retaining the independence to choose which eligible vehicles and verified technologies best meet their communities' needs. The language in

Diesel Exhaust Health Effects

Direct emissions from diesel engines, especially $PM_{2.5}$, NO_x, and sulfur oxides (SO_x), contribute to health problems. In addition, NO_x contributes to the formation of ozone and PM through chemical reactions.

$PM_{2.5}$ has been associated with an increased risk of premature mortality, increased hospital admissions for heart and lung disease, and increased respiratory symptoms. Long-term exposure to components of diesel exhaust, including diesel PM and diesel exhaust organic gases, are likely to pose a lung cancer hazard. Exposure to ozone can aggravate asthma and other respiratory symptoms, leading to more asthma attacks, the use of additional medication, more severe symptoms that require a doctor's attention, more lost school and work days, more visits to the emergency room, increased hospitalizations, and even premature mortality. People in many areas of the United States experience short-term (one to three hours) and prolonged ozone exposures (six to eight hours), which have been linked to diminished lung function, greater respiratory symptoms, and increased hospital visits. Repeated exposure to ozone can increase susceptibility to respiratory infection and lung inflammation and can aggravate preexisting asthma. At sufficient concentrations, ozone can even cause permanent damage to the lungs, including the development of chronic respiratory illnesses. Children, outdoor workers, those who exercise outdoors, people with heart and lung disease, and the elderly are most at risk.

The technologies used in DERA grants can reduce PM emissions by up to 95 percent, HC and NO_x by up to 90 percent, and greenhouse gases by up to 20 percent. Each of these reductions makes an immediate and positive impact on public health.

the reauthorization of DERA also allows DERA funds to be used for local and state mandates. In addition, funds may be used for measures included under State Implementation Plans (SIPs).

Approximately 70% of competitive projects have taken place in nonattainment areas for PM$_{2.5}$ (per the 2006 standard) or 8-hour ozone (per the 2008 standard).[6] In addition, most projects initiated in areas of attainment address localized areas of pollution such as near ports, railyards, or bus depots to benefit those disproportionately affected by diesel exhaust.

Fostered Clean Diesel Projects on Tribal Lands, Territories, and Islands

Tribal grants are another important component of the program's commitment to environmental justice. EPA has awarded more than $2 million for eight tribal grants in Minnesota, California, Alaska, Arizona, Washington, and Iowa. The tribal projects have retrofitted or replaced school buses, refuse haulers, transport buses, utility vehicles, fishing vessels, and mining equipment.

DERA began funding projects in the District of Columbia in FY 2009, adding Puerto Rico in FY 2011. DERA's most recent authorization that took effect on October 1, 2011, allows Guam, the United States Virgin Islands, American Samoa, and the Commonwealth of the Northern Mariana Islands to receive federal funds for clean diesel projects.

Figure 3. Applications and Funding Requested vs. Awarded for the DERA National Competition

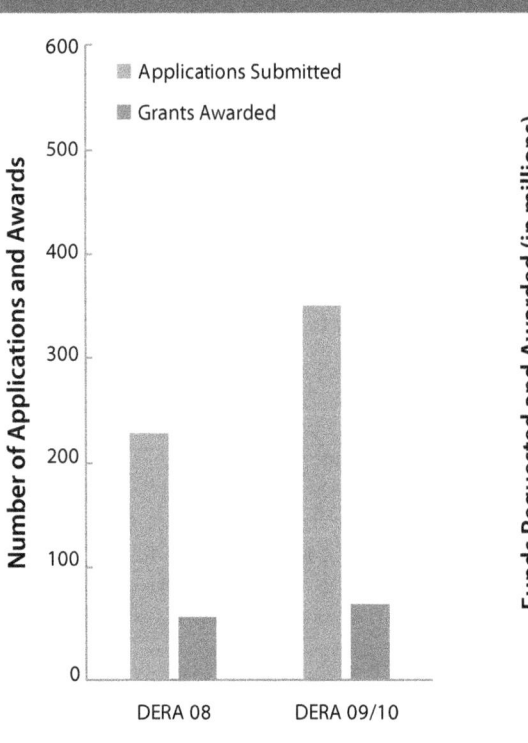

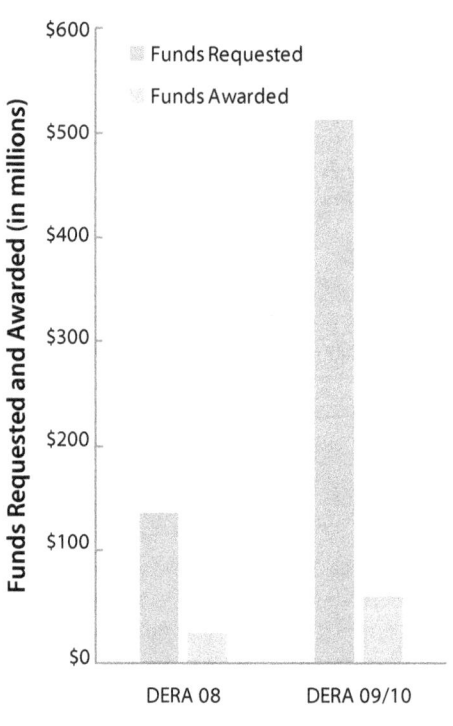

---

[6] The percentage of projects taking place in 2011 in non-attainment areas was calculated using the EPA Office of Air and Radiation's most recent National Ambient Air Quality Standards, which can be found at www.epa.gov/air/criteria.html.

Tribal Communities Use EPA Grants to Support Clean Diesel Projects

EPA awarded eight tribal grants totaling more than $2 million in FY 2009/2010 and FY 2011. In the months prior to the FY 2009/2010 Request for Proposals, EPA encouraged tribal participation by attending conferences and reaching out to eligible entities. Tribal or tribal coalitions with jurisdiction over transportation or air quality could apply on their own or partner with nonprofit organizations that work with diesel fleet operators to reduce pollution. These projects retrofitted or repowered school buses, utility vehicles, mining equipment, refuse haulers, fishing vessels, and a generator in eight tribal communities in Minnesota, California, Arizona, Washington, Iowa, and Alaska.

Soboba Band of Luiseño Indians Retrofit School Buses

$78,000 DERA 2009/2010 grant
$18,300 in matching funds

The Soboba Band of Luiseño Indians, located in Riverside, California, received a $78,000 FY 2009/2010 grant to retrofit six school buses. The diesel particulate filters they installed will reduce CO and PM emissions by 90 percent annually.

Buses in this fleet travel nearly 700 miles per day and serve children at 16 reservations in southern California. The buses travel through Riverside, San Diego, and San Bernardino counties. All three are heavily populated, and Riverside and San Bernardino are among the areas of the country with the highest air pollution levels. In addition, the Soboba Tribe hopes to maximize the impact of this project by educating other tribes on the importance of using this technology and partnering with them in similar efforts.

DERA Program Structure Reflects National and State Interests

The DERA program has two parts: a state allocation component, titled the State Clean Diesel Grant Program, and a national program. The state program has received 30 percent of the total funding. The national program has received the remaining 70 percent and consists of three separate competitive programs:

➲ The National Clean Diesel Funding Assistance Program

➲ The National Clean Diesel Emerging Technologies Program

➲ The SmartWay Clean Diesel Finance Program

Components of the National Clean Diesel Campaign

National Clean Diesel Funding Assistance Program

Awarded approximately $250 million in grants, including funds to tribal nations

Seven regional collaboratives administer the grant competition

Dedicated to deploying verified and certified technologies

Clean Diesel Emerging Technologies Program

Awarded approximately $30 million in grants

Provides opportunities to advance cutting-edge technologies

National Clean Diesel Program

SmartWay Finance Clean Diesel Program

Awarded approximately $50 million in grants

Establishes innovative finance models to provide funding to fleets

State Grant Program

Awarded approximately $130 million in grants to all 50 states, the District of Columbia, and island territories

Direct funding assistance to states for diesel emissions reduction sub-grants and loans

Regional Clean Diesel Collaboratives Have Fostered Local Approaches

Reducing diesel emissions is a shared responsibility, and EPA encourages collaboration among key stakeholders from state and local government agencies (including EPA Regional Offices), environmental and community organizations, fleet owners/operators, private industry, and others to pool talent and resources to achieve better air quality through clean diesel initiatives. These seven regional Clean Diesel Collaboratives are diverse, multi-stakeholder groups that leverage funds, provide technical assistance, nurture partnerships, and pursue local approaches to mitigating diesel emissions.

By linking regional stakeholders and coordinating efforts, the Collaboratives are able to achieve significant emissions reductions across large geographic areas (see Figure 4, "Regional Clean Diesel Collaboratives"). For more information about the Collaboratives, please see Appendix F.

Funding for States and U.S. Territories

The State Clean Diesel Grant Program is a formula allocation program in which all 50 states are eligible. The District of Columbia became eligible beginning in FY 2009 and Puerto Rico in FY 2011. The DERA Reauthorization signed in January 2011, authorizes Guam, the United States Virgin Islands, American Samoa, and the Commonwealth of the Northern Mariana Islands to receive funding. Thirty percent of appropriated DERA funds must be allocated to the states to implement grant and loan programs for clean diesel projects. From 2008 to 2010, the states received $133.6 million through this program.

Figure 4. Regional Clean Diesel Collaboratives

DERA TOOLS AND RESOURCES

EPA has taken steps to expand the adoption of clean diesel technologies. To this end, EPA has provided informative and useful tools to the general public and diesel equipment operators and will continue to engage directly with transport companies across the country in a voluntary initiative to reduce their use of diesel fuel. EPA has also helped manufacturers introduce new clean diesel technologies to the marketplace.

Emerging Technologies Program

Since its formation, the Emerging Technologies (ET) program has played an important role in encouraging the development of cutting-edge pollution reduction solutions and bringing the next generation of emissions-reducing technologies to market. This program has expedited technologies under development through the verification or certification process so that they can be quickly and widely adopted by fleets in marine, locomotive, nonroad, and highway applications around the country.

The ET program has served as an initial step for technologies to obtain full verification. To date, more than 10 different emerging technology participants have benefitted from partnering with fleets to demonstrate and improve 17 different technologies.

To read more about the ET program and other clean diesel technologies, please visit www.epa.gov/cleandiesel/verification/verif-list.htm or www.epa.gov/cleandiesel/verification/emerg-list.htm.

SmartWay Finance Program

The SmartWay Finance Clean Diesel Program is an NCDC grant competition that has awarded DERA funds for innovative finance programs to provide financial incentives, such as low-cost loans and loan guarantees, to fleet owners for the purchase of fuel-saving and emission control technologies and vehicle replacements. Since 2008, grants totaling almost $50 million have been awarded to organizations to establish financing programs that assist small and medium-sized fleet owners in purchasing cleaner, more fuel-efficient trucks and equipment.

Additionally, the SmartWay technology program is closely integrated with DERA's verification efforts. Technical staff has evaluated the efficiency and emissions performance of technical and operational strategies with a focus on idle reduction, truck and trailer aerodynamic components, low-rolling-resistance tires, and retrofit technologies. Vehicles including tractor-trailers that meet performance criteria have earned SmartWay designation.

For more information about SmartWay, visit www.epa.gov/smartway.

Public Outreach

EPA offers the public technical tools to learn about and explore environmental issues and solutions, including clean diesel activities, in their area. One such tool is a new interactive application called *MyEnvironment*, which displays outcomes of clean diesel projects as well as information about organizations, technologies, and health benefits associated with the projects. For more information, visit www. epa.gov/myenvironment.

Additionally, NCDC expanded on past work with school bus fleets and Scholastic, Inc., by releasing a Spanish translation of *The Magic School Bus Gets Cleaned Up (El Autobús Mágico Necesita una Limpieza.)* The book's launch was held in Biloxi, Mississippi, in March 2012. More than 1,000 children from across the state were in attendance to participate in interactive activities to learn about environmental issues including air pollution and celebrate the book's release.

Mississippi Cleans Its Entire Fleet of School Buses

The state of Mississippi has cleaned its entire fleet of eligible school buses, and now 100 percent of these buses meet the newest emission standards or are equipped with emission control devices. Bolstered by DERA funding, Mississippi has raised additional money to replace 52 buses and to equip 2,000 more with emissions-reducing technologies. Nearly 500,000 children in Mississippi benefit each day from riding clean school buses. In addition, Mississippi counties along the Gulf of Mexico have implemented an idle education program, for which several schools conduct studies to evaluate the amount of idling that occurs around school grounds in an effort to further reduce children's exposure to diesel exhaust.

2. Highlights of the DERA 2008 and 2009/2010 Grants

I n every state, some Americans now breathe cleaner air thanks to funding awarded through the Diesel Emissions Reduction Program. DERA grants have had an immediate and positive impact on public health and air quality. The initial funding in 2008 established a firm foundation for the program and FY 2009/2010 grants fine-tuned the award process to increase overall cost-effectiveness.

2008: DERA's Debut

In FY 2008, the DERA program made its debut. In its first grant cycle, EPA received four times as many applications as it could fund, ensuring that the first round of projects would be strong candidates for success.

With a total of nearly $50 million awarded to establish 119 projects, grantees upgraded more than 14,000 vehicles or pieces of equipment. These projects reduced 22,700 tons of NO_x, 2,700 tons of PM, 4,200 tons of HC, 15,900 tons of CO, and 289,900 tons of CO_2. The 2008 grants provided approximately $644 million to $1.6 billion in health benefits and saved nearly 26 million gallons of fuel.

For a complete list of grants, including sector, technology, and total funding, please see Appendix A.

Figure 5. DERA 2008 Grants: Retrofitted or Replaced Diesel Engines

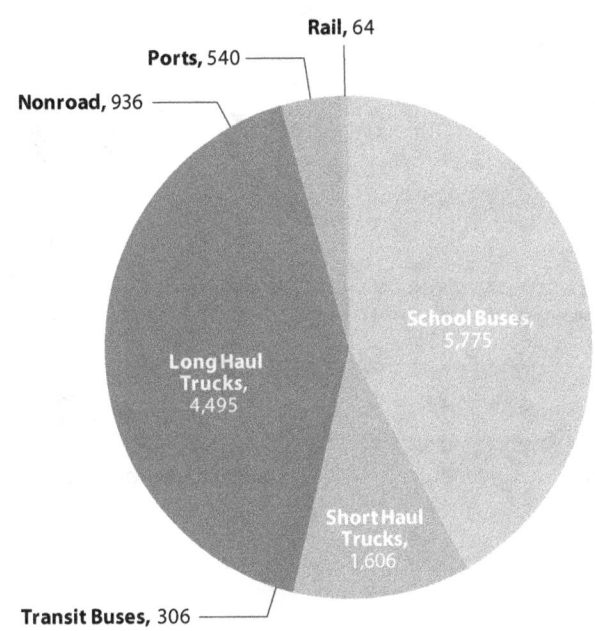

Rail, 64
Ports, 540
Nonroad, 936
School Buses, 5,775
Long Haul Trucks, 4,495
Short Haul Trucks, 1,606
Transit Buses, 306

Figure 6. Technologies Used in the FY 2008 DERA Grants

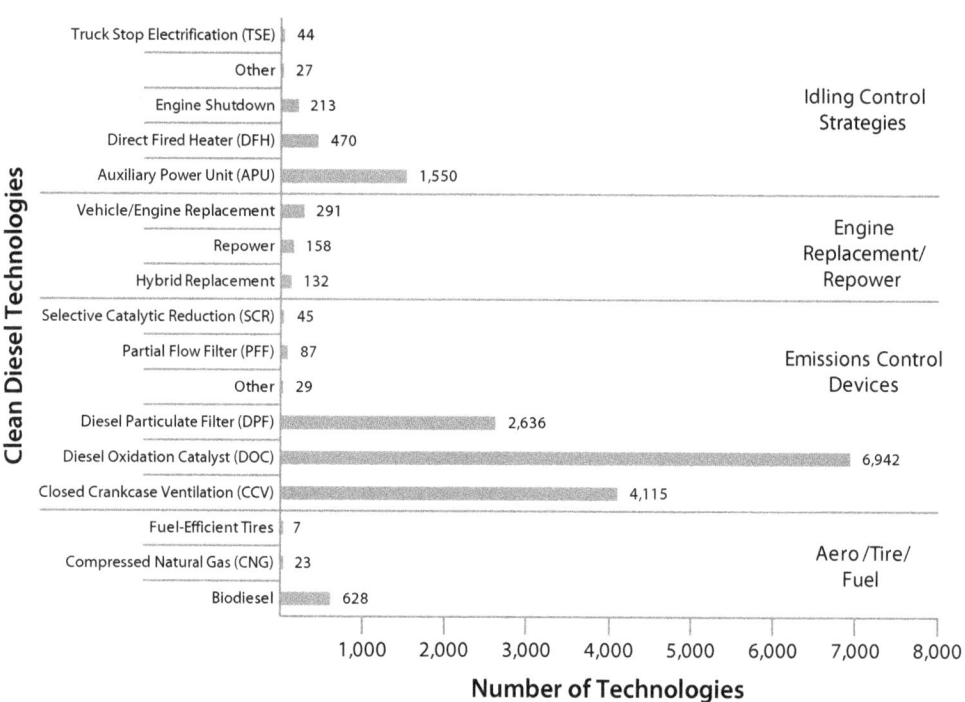

Idling Control Strategies:
- Truck Stop Electrification (TSE): 44
- Other: 27
- Engine Shutdown: 213
- Direct Fired Heater (DFH): 470
- Auxiliary Power Unit (APU): 1,550

Engine Replacement/Repower:
- Vehicle/Engine Replacement: 291
- Repower: 158
- Hybrid Replacement: 132

Emissions Control Devices:
- Selective Catalytic Reduction (SCR): 45
- Partial Flow Filter (PFF): 87
- Other: 29
- Diesel Particulate Filter (DPF): 2,636
- Diesel Oxidation Catalyst (DOC): 6,942
- Closed Crankcase Ventilation (CCV): 4,115

Aero/Tire/Fuel:
- Fuel-Efficient Tires: 7
- Compressed Natural Gas (CNG): 23
- Biodiesel: 628

Y-axis: Clean Diesel Technologies

X-axis: Number of Technologies (1,000 – 8,000)

Montgomery County Children Ride Clean Buses to School

DERA in Action

$700,000 DERA FY 2008 grant

The Montgomery County (Maryland) public schools received almost $700,000 in 2008 to reduce emissions from their school bus fleet. Montgomery County used those funds to equip 86 buses with diesel particulate filters. These filters reduce PM emissions by over 85 percent, which is especially important for children in Montgomery County with asthma who ride the buses every day. By reducing pollution from its school buses, Montgomery County has healthier students and cleaner air for all its residents.

Miami Replaces Refuse Haulers With Cleaner, More Fuel-Efficient Model

$700,000 DERA grant
$2.1 million in matching funds

The city of Miami, Florida, received approximately $700,000 in DERA funding in 2009 to replace 10 refuse haulers with cleaner, more fuel-efficient models, including four early model hydraulic hybrids. These hybrids use regenerative braking to save fuel by as much as 30 percent. The fuel savings reduce CO_2 emissions by approximately 30 tons per vehicle per year. Other benefits include less brake wear, lower engine noise, and overall lower operating cost.

Hydraulic hybrid refuse haulers can operate at low speeds (zero to about 30 miles per hour) on the hydraulic system without using any diesel fuel. The hydraulic system is recharged by braking and, given the typical route of a refuse hauler, there are usually many start-and-stops throughout the work day.

These fuel-efficient hydraulic hybrid refuse haulers now operate throughout Miami.

Government Partnership in Missouri Cuts Pollution and Reduces Fuel Consumption

$725,000 DERA FY 2008 grant
$1.2 million in matching funds

The Missouri Department of Transportation teamed up with the Missouri Department of Natural Resources to upgrade 132 municipal vehicles. They received a $725,000 DERA grant in 2008 and raised more than a million dollars in matching funds to accomplish their goal and reduce diesel exhaust. Throughout this project, two state agencies successfully worked together to promote idle reduction and other emissions-reducing technologies throughout the state. Through implementation of this project, over 1,000 tons of diesel emissions will be reduced from the project vehicles over their lifetimes.

DERA in Action

A Recovery Act grant to Minnesota's Project Green Fleet funded APUs for two Twin Cities & Western Railroad Company trains to reduce pollution from idling.

2009/2010: CREATING EFFICIENCIES

In both FY 2009 and FY 2010, the DERA program received an appropriation of $60 million, which EPA combined to create a single funding competition of $120 million to establish clean diesel projects. Combining the two award years streamlined the Request for Proposals process and gave applicants the opportunity to propose larger, more impactful projects.

For FY 2009/2010, EPA received over 350 applications and the funding requested outweighed available dollars by five to one. In late 2010, EPA awarded 84 grants totaling $120 million.

These grants will provide estimated lifetime emissions reductions of 50,600 tons of NO_x, 2,600 tons of PM, 3,600 tons of HC, 9,300 tons of CO, and 706,000 tons of CO_2 with fuel savings of nearly 63 million gallons. These grants provide lifetime health benefits of $728 million to $1.8 billion.

For a complete list of grants, including sector, technology, and total funding, please see Appendix A.

Figure 7. DERA 2009/2010 Grants: Retrofitted or Replaced Diesel Engines

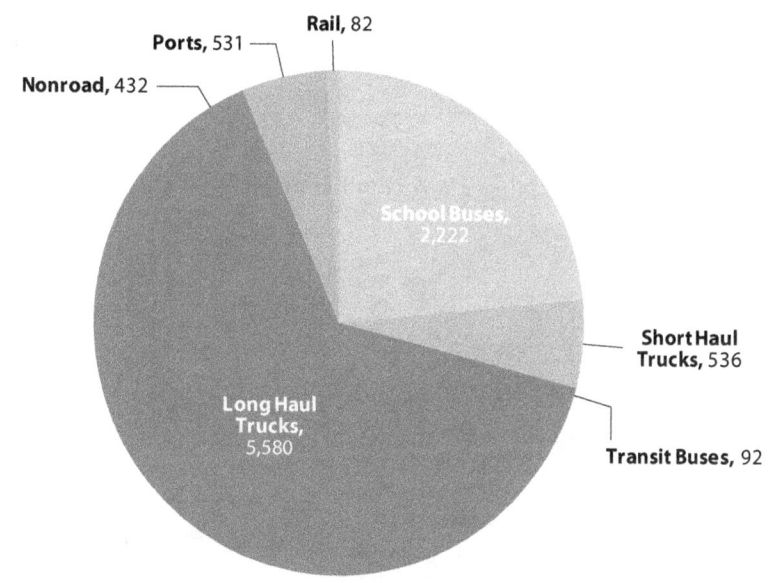

More applicants requested grants for repower and replacement projects in FY 2009/2010 than in FY 2008. DERA funding for replacement and repowers is especially helpful to small business owners because it makes vehicle upgrades and emissions reductions affordable. Many of these replacements, especially in the marine, port, and rail sectors, take place in environmental justice areas and create emissions reductions for multiple pollutants—lowering NO_x, PM, HC, CO, and CO_2.

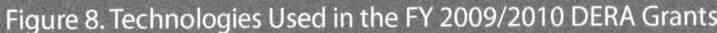

Figure 8. Technologies Used in the FY 2009/2010 DERA Grants

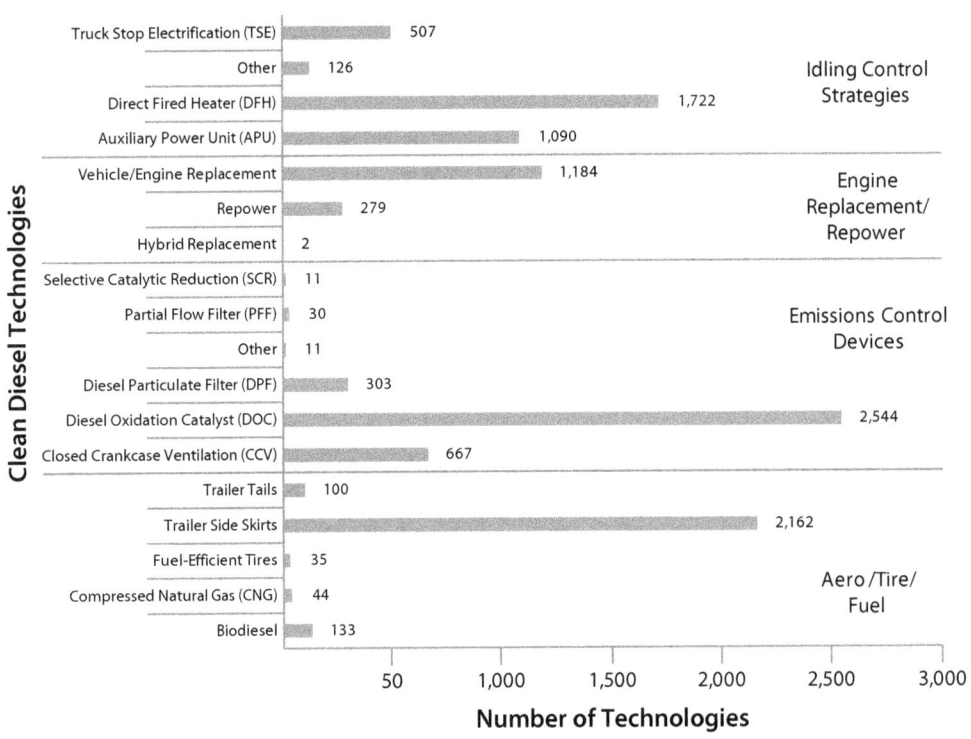

Clean Diesel Technologies (y-axis)

Idling Control Strategies
- Truck Stop Electrification (TSE): 507
- Other: 126
- Direct Fired Heater (DFH): 1,722
- Auxiliary Power Unit (APU): 1,090

Engine Replacement/Repower
- Vehicle/Engine Replacement: 1,184
- Repower: 279
- Hybrid Replacement: 2

Emissions Control Devices
- Selective Catalytic Reduction (SCR): 11
- Partial Flow Filter (PFF): 30
- Other: 11
- Diesel Particulate Filter (DPF): 303
- Diesel Oxidation Catalyst (DOC): 2,544
- Closed Crankcase Ventilation (CCV): 667

Aero/Tire/Fuel
- Trailer Tails: 100
- Trailer Side Skirts: 2,162
- Fuel-Efficient Tires: 35
- Compressed Natural Gas (CNG): 44
- Biodiesel: 133

x-axis: 50, 1,000, 1,500, 2,000, 2,500, 3,000

Number of Technologies

California Repowers Switcher Locomotives to Clean Rail Yards

DERA in Action

$3,949,500 DERA 2009/2010 grant
$1,050,500 in matching funds

The California Air Resources Board received $3,949,496 in FY 2009/2010 to repower four aging locomotives that operate in the South Coast, the San Joaquin Valley, and the San Francisco Air Basins with modern Tier 3 engines.

These locomotives, also known as generator set switch locomotives, can have an operational lifetime of more than 30 years. Switch locomotives typically operate in and around rail yards to put trains together and move railcars between rail yards.

Each repowered locomotive can reduce NO_x emissions by about 235 tons and PM emissions by 11 tons.

NESCAUM's Tower Gantry Crane Repower Project in the New York Metropolitan Area

DERA in Action

$1,420,000 DERA 2009/2010 grant
$408,000 in matching funds

Gantry cranes are used in every high-rise construction project and are the most visible pieces of equipment on any construction site. Their engines (and other heavy-duty diesel engines) emit a significant amount of diesel emissions into the air, contributing approximately 9 percent and 11 percent of the New York Metropolitan Area's total NO_x and PM emissions, respectively.[7]

The Northeast States for Coordinated Air Use Management (NESCAUM), partnered with Cornell & Company, Inc., is repowering 17 tower gantry cranes equipped with unregulated engines with newer, cleaner EPA-certified engines.

Two Cornell Tower Gantry Cranes operating in Lower Manhattan at the site of Three World Trade Center.

The gantry cranes targeted by this project are most frequently operated in urban centers to help build commercial and residential skyscrapers. Emissions from major urban construction projects can make poor air quality even worse for local, highly dense populations.

This project will significantly reduce emissions of NO_x and diesel PM within New York and New Jersey, with a secondary benefit to Pennsylvania and other areas outside the New York metro area. Cornell & Company, Inc., Equipment Manager Donald Garrahan stated, "We have served the Northeast USA for many years and are enthusiastic about the prospect of improving air quality and living conditions in our home marketplace."

---

[7] Calculated by NESCAUM using the 2005 National Inventory Emissions Database.

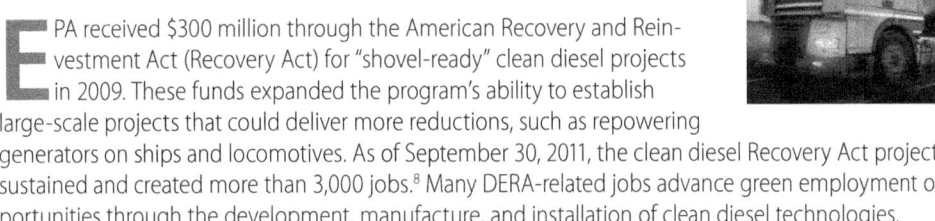

3. American Recovery and Reinvestment Act: A Boost for DERA

EPA received $300 million through the American Recovery and Reinvestment Act (Recovery Act) for "shovel-ready" clean diesel projects in 2009. These funds expanded the program's ability to establish large-scale projects that could deliver more reductions, such as repowering generators on ships and locomotives. As of September 30, 2011, the clean diesel Recovery Act projects sustained and created more than 3,000 jobs.[8] Many DERA-related jobs advance green employment opportunities through the development, manufacture, and installation of clean diesel technologies.

More than 600 applicants seeking $1.7 billion responded to the program's Request for Proposals for Recovery Act funding—nearly seven applications for every one awarded, requesting $10 for every dollar available. These applicants proposed $2.2 billion in matching funds. In mid-2009, EPA awarded 160 projects across the country.

For a complete list of Recovery Act grants, please see Appendix A.

In total, the Recovery Act projects save more than 116 million gallons of fuel. They also reduce approximately 130,600 lifetime tons of NO_x, 7,200 tons of PM, 10,200 tons of HC, 22,800 tons of CO, and 1,309,100 tons of CO_2, creating health benefits of $2 billion to $4.9 billion.

Once EPA received the funds, it acted quickly to competitively award the funding and establish 160 new projects nationwide. On March 27, 2009, EPA awarded its first Recovery Act grant under the DERA program to the state of Colorado. Colorado received $1.73 million to retrofit nearly 1,000 school buses, creating service-related jobs and reducing Colorado's children's exposure to diesel exhaust.

NESCAUM received a $1.65 million Recovery Act grant and used part of it to repower this New Hampshire sightseeing boat's engine.

---

[8] This jobs estimate was created based on self-reported information from Recovery Act grant recipients according to the Office of Management and Budget's guidance on job reporting.

Figure 9. Vehicles Retrofitted With Recovery Act Grants

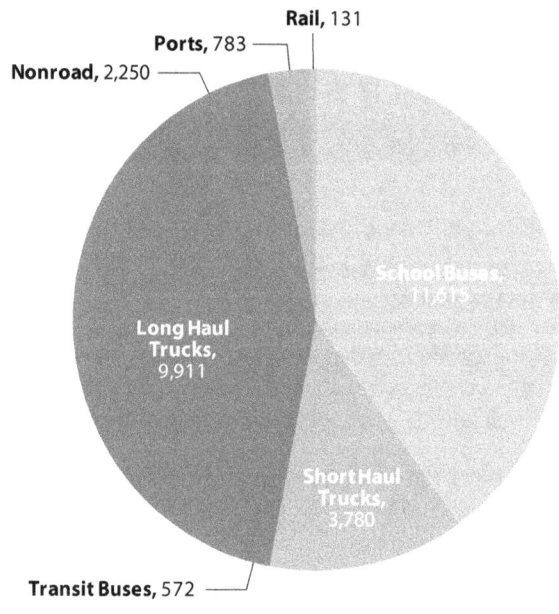

Rail, 131
Ports, 783
Nonroad, 2,250
School Buses, 11,615
Long Haul Trucks, 9,911
Short Haul Trucks, 3,780
Transit Buses, 572

Portland, Oregon, Cleans Up Construction Equipment

$400,000 DERA 2008 grant
$1.5 million DERA Recovery Act grant
$54,000 in matching funds

DERA in Action

EPA awarded two grants totaling about $2 million to reduce diesel pollution in the Portland area: a grant in FY 2008 for approximately $400,000 to the Portland-Multnomah Clean Diesel Partnership and a Recovery Act grant in 2009 for more than $1.5 million to the City of Portland Clean Diesel Partnership. The City of Portland, the Oregon Department of Environmental Quality, and Multnomah County provided matching funds for the projects. These grants supported clean diesel technology devices on public fleets and construction equipment operating in environmental justice neighborhoods.

The Partnership Project has also implemented an idle reduction policy and a pilot program to hire contractors who use clean diesel equipment. Today, the City of Portland Clean Diesel Partnership Project is well on its way to achieving regional goals to reduce emissions from public fleets and construction equipment.

Figure 10. Technologies Used in the Recovery Act Grants

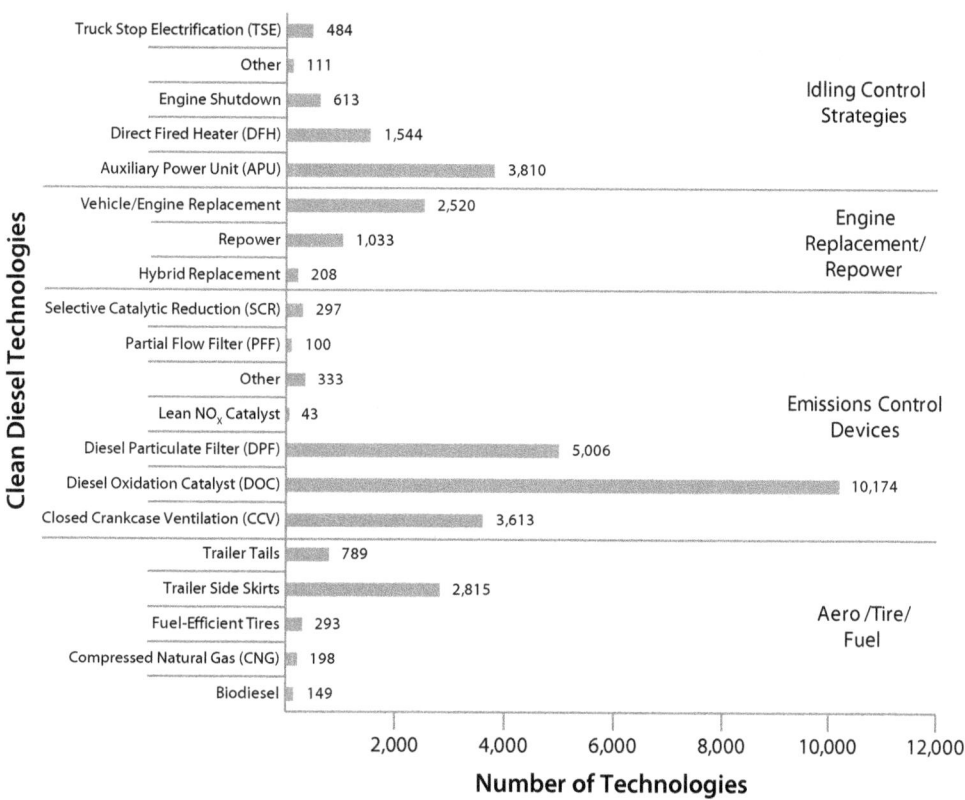

Clean Diesel Technologies

Technology	Value
Truck Stop Electrification (TSE)	484
Other	111
Engine Shutdown	613
Direct Fired Heater (DFH)	1,544
Auxiliary Power Unit (APU)	3,810
Vehicle/Engine Replacement	2,520
Repower	1,033
Hybrid Replacement	208
Selective Catalytic Reduction (SCR)	297
Partial Flow Filter (PFF)	100
Other	333
Lean NO_x Catalyst	43
Diesel Particulate Filter (DPF)	5,006
Diesel Oxidation Catalyst (DOC)	10,174
Closed Crankcase Ventilation (CCV)	3,613
Trailer Tails	789
Trailer Side Skirts	2,815
Fuel-Efficient Tires	293
Compressed Natural Gas (CNG)	198
Biodiesel	149

Idling Control Strategies

Engine Replacement/ Repower

Emissions Control Devices

Aero/Tire/ Fuel

Number of Technologies

Chesapeake Bay Repowers Old Engines

$1,300,000 DERA Recovery Act grant
$747,000 in matching funds

The Chesapeake Bay Foundation (CBF) is a nonprofit organization dedicated to improving and maintaining the environmental quality of the Chesapeake Bay and its rivers and streams. Nitrogen is the number one pollutant of the Bay's watershed, and much of the air pollution comes from NO_x emissions.

One of CBF's strategies to reduce nitrogen pollution is to focus on cleaning up operations of the marine community. Large commercial tug boats, local fishing vessels, and other recreational and commercial boats regularly traverse the Bay, and repowering older diesel engines with newer, cleaner ones can reduce NO_x, reducing the amount of nitrogen that could be absorbed from the air by the waterbody. To assist with this effort, EPA awarded CBF a Recovery Act grant of $1.3 million. CBF provided an additional $547,294 and leveraged $200,000 more to purchase and install new engines in six CBF education vessels, four working boats, and one tug boat.

This project has benefitted a wide range of boat users. For example, CBF uses the six re-powered education vessels as floating classrooms where instructors teach students (grades 6–12), teachers, principals, and other school administrators about the Bay's unique ecology. The repowered engines on the education vessels provide a learning opportunity to demonstrate the connection between air pollution and water pollution for students and teachers. CBF has also partnered with Tangier Sound commercial watermen and ferry and charter boat captains to repower their workboats with cleaner engines that reduce diesel emissions. Workboats are used for commercial and recreational fishing and as passenger ferries for excursions and fishing charters. The repowered engines in these workboats help reduce fuel usage and air emissions.

Minnesota's Project Green Fleet Helps Clear the Air

$3,000,000 DERA Recovery Act grant
$647,800 in matching funds

Project Green Fleet works to help Minnesota avoid a non-attainment designation for its air quality. Part of this project involves installing diesel oxidation catalysts, diesel particulate filters, auxiliary power units, and other idle reduction technologies on everything from school buses and municipal vehicles to construction equipment and trains.

Photo courtesy of the Blue Cross and Blue Shield of Minnesota Foundation.

In July 2009, EPA awarded a Recovery Act grant for $3 million to support Project Green Fleet to retrofit nearly 600 pieces of on-road and off-road equipment in Minnesota. These upgrades included the installation of 390 pieces of exhaust control equipment, 200 pieces of idle reduction equipment, 25 engine repowers, and two vehicle replacements. In addition, the project implemented a statewide outreach and communications strategy to increase public awareness about the negative environmental and health consequences of diesel pollution.

Project Green Fleet has reduced NO_x emissions by more than 180 tons and PM by 13 tons annually. Today, thanks to this innovative program, Minnesotans are breathing easier and enjoying a healthier environment, and they are more aware of efforts to reduce diesel engine pollution.

4. Looking to the Future

CHALLENGES AND LESSONS LEARNED

EPA has been successful in providing immediate health benefits by reducing diesel emissions in a cost-effective manner. With more than three years of experience awarding grants in this sector, EPA has encountered and learned from some challenges.

Effective Management and Tracking of Grants Helps Projects Meet Goals

For each award competition, EPA has employed rigorous selection criteria. The Requests for Proposals that EPA has issued require very detailed information from applicants about their project plans, budgets, and estimated emissions reductions. Once grants are awarded, the recipient must submit quarterly progress reports, and Recovery Act grant recipients submit online reports for job tracking purposes. Additionally, EPA has used audits to monitor grant progress. About a quarter of grants received onsite audits in 2010.

EPA also tracks projects in the Database for Reporting Innovative Emissions Reductions (DRIVER), which provides real-time funding information. In this manner, grant coordinators are able to easily track the more than 500 DERA-funded projects, plus another 1,000 clean diesel projects nationwide.

EPA's Office of Inspector General (OIG) conducted one audit on the DERA program in March 2010 and two in March 2011.[9] The OIG found program strengths in the award process and staff knowledge about the grants, but suggested areas for improvement, including a need for additional technical guidance for grant managers and recipients. The DERA staff implemented these recommendations to manage grants more effectively, such as enhanced training for EPA grant project officers and grantees. EPA also provided additional training and information on clean diesel technologies.

---

[9] The March 2010 report can be found at www.epa.gov/oig/reports/2010/20100323-10-R-0082.pdf. The two March 2011 reports can be found at www.recovery.gov/Accountability/inspectors/Documents/20110301-11-R-0141_ARRA.pdf and www.recovery.gov/Accountability/inspectors/Documents/20110328-11-R-0179_ARRA.pdf.

Product Delivery Delays May Prevent Grantees From Reaching Their Project Deadlines

At times, grantees have encountered delays in product delivery, which has in turn delayed project implementation. For example, grant recipients placed a large number of Recovery Act orders to product vendors in a short time frame, which created a bottleneck in the processing and delivery of technology and equipment. Furthermore, some materials are used in multiple verified and emerging technologies, so unforeseen manufacturing delays affected a variety of clean diesel projects.

The grantees' progress is tied to the volume of technology orders and vendor capacity. To help grantees overcome this obstacle, in situations where delays are unavoidable and out of the grant recipient's control, EPA offered a no-cost extension so grantees could complete the project. In addition, to prevent product back orders and delays, EPA maintains regular communication with national vendors regarding their supply and notifies the Regional Collaboratives when such problems do occur for grant planning purposes.

Macroeconomic Trends Impact Project Partners

Sometimes macroeconomic trends have an unforeseen impact on project planning and progress. Over the past few years, EPA witnessed how the economic downturn prevented some grant recipients' project partners from finding fleets and generating matching funds. This caused delays in the projects while grantees sought new partners.

To hedge against the impacts of larger economic downturns, EPA now encourages grant applicants and recipients to identify multiple potential project partners in case one is unable to participate.

The San Diego County Air Pollution Control District received a $1.6 million Recovery Act grant to replace or retrofit over 125 buses.

Clean Diesel Opportunities

The DERA program has been successful in reducing pollution from the legacy fleet of diesel engines, and recent funding competitions continue to generate applications. For FY 2011, EPA received 179 applications requesting $312 million, or $6 for every one EPA had available. EPA awarded 48 competitive grants, in addition to the state awards. For FY 2012 EPA received 93 applications requesting $131 million, or approximately $5 for every one EPA had available. These projects will continue to chip away at diesel emissions from the legacy fleet.

Virginia Clean Cities received a $1 million Recovery Act grant from the DERA program. This money funded four projects throughout Virginia that retrofitted school buses, transit buses, and refuse haulers.

Fiscal Year 2012 and Beyond

Almost four years of DERA implementation have provided EPA with information on more than 500 grants across the country. Analysis of the results of these grants has allowed EPA to more finely hone criteria for evaluating and assessing clean diesel project applications. Factors such as useful life of the engines, model year, vehicle-miles-traveled, cost, type and quantity of reductions of pollution, technology specifications, and location all play a role in the effectiveness of any given project. More specific evaluation criteria ensure that projects are the most effective and conducted in the areas of greatest need.

EPA awarded $30 million in FY 2012 for new clean diesel projects. Under the national grant competition, EPA refined its Request for Proposals to encourage the most cost-effective and impactful projects. EPA assigned points to project proposals taking place in locations that have been designated nonattainment areas or in places that receive a disproportionate amount of diesel emissions, such as ports and railyards. EPA also assigned points to encourage participants to submit proposals that included certain model year vehicles paired with verified technology to maximize emission reductions over the lifetimes of the vehicles.

In January 2011, DERA was reauthorized through 2016, allowing up to $100 million in appropriations for each fiscal year. DERA's reauthorization also allows for a rebate program, thus diversifying the types of financial tools that EPA has to reduce emissions from diesel vehicles. Rebates could streamline the funding process, providing immediate financial incentives directly to fleets that retrofit and replace older diesel engines. In fall 2012, EPA plans to pilot this concept by offering $2 million in rebates to school bus fleet owners. EPA will collect applications and randomly select recipients from the pool of eligible applicants and vehicles. Limiting eligibility to specific model years, requiring the scrappage of old vehicles, and other requirements will apply. Any future rebate program could focus on other diesel emission reduction strategies, fleets, or locations.

The DERA program has shown that retrofits and engine replacements are effective in reducing emissions and provided valuable lessons in how to administer clean diesel programs. Going forward, EPA plans to sharpen its focus on any remaining areas of disproportionate exposure to emissions from diesel engines, and ensure that clean diesel projects are as cost-effective as possible. EPA recognizes the limited availability of federal funding and has proposed to transition the program to greater reliance on state and local efforts to address diesel emissions from legacy fleets.

In FY 2013 the Agency will pilot a new approach that will target specific fleets in high diesel exposure areas such as near ports and freight distribution hubs and other disproportionately affected communities. The new strategy would allocate funds to a new rebate program established under DERA's reauthorization, and to grants, in part, for finance mechanisms that help fleets reduce diesel emissions. EPA believes the rebate, grant, and finance programs may allow greater precision in scrapping certain model years of vehicles and equipment and assisting public and private fleet owners with retrofitting or replacing those engines.

As the program looks ahead to the challenges of cleaner movement of goods through the nation's supply chain, reducing black carbon pollution, and assisting environmentally challenged communities, DERA will continue to follow its guiding principles for all future implementation:

- ➲ Target areas and populations that are disproportionately affected by diesel emissions.

- ➲ Continue to reduce pollution from diesel engines by partnering with key stakeholders.

- ➲ Provide assistance to state and local governments in the development of their own clean diesel programs.

- ➲ Continue working with states to provide numerous and effective clean diesel technology options to project partners.

- ➲ Continue confirming emission performance of verified technologies in the field.

- ➲ Maximize health benefits from clean diesel projects.

Appendix A: National Clean Diesel Funding Assistance Program

FY 2008 Grants					
State	Grant Recipient	EPA Grant Amount*	Match	Project Target Fleet(s)	Technology Type(s)
CA	CALSTART, Inc.	$678,459		Construction, Delivery Truck, Transit Buses	Retrofits
CA	Kern County Superintendents of Schools	$540,000		School Bus	Idle Reduction, Replacement/ Repower, Retrofits
CA	Sacramento Metropolitan Air Quality Management District	$553,360		Long Haul Trucks, School Bus, Short Haul	Retrofits
CA	South Coast Air Quality Management District	$1,000,000		Long Haul Trucks	Retrofits
CO	City and County of Denver	$178,183	$87,059	Refuse Hauler, Utility Vehicle	Idle Reduction
CO	Colorado Department of Public Health	$399,999		Construction, School Bus	Idle Reduction, Replacement/ Repower, Retrofits
CO	Regional Air Quality Council	$455,645		Long Haul Trucks, Utility Vehicle	Idle Reduction
CT	Connecticut Department of Environmental Protection	$49,867		Construction, Delivery Truck, Short Haul	Retrofits
CT, MA, ME, NH, RI	Environmental Defense Fund	$400,000	$2,025,000	City/County Vehicle, Utility Vehicle	Retrofits
CT, MA, ME, NH, RI	Northeast States for Coordinated Air Use Management	$319,301	$100,000	Construction	Retrofits
DC	Metropolitan Washington Council of Governments	$486,866		Construction	Replacement/Repower, Retrofits
FL	City of St. Petersburg	$396,709	$5,745	Construction, Delivery Truck, Short Haul	Clean/Alt Fuels
GA	Georgia Ports Authority	$250,000	$33,075	Marine, Ports and Airports	Retrofits
ID	Idaho Department of Environmental Quality	$481,303		School Bus	Retrofits
IL	Illinois Environmental Protection Agency	$678,604		Delivery Truck, Long Haul Trucks, Refuse Hauler, School Bus, Short Haul, Transit Buses	Idle Reduction, Replacement/ Repower, Retrofits
IN	Elkton Pigeon Bay/Laker School District	$251,100		School Bus	Clean/Alt Fuels, Idle Reduction, Retrofits
IN	Indiana Department of Environmental Management	$334,500		Long Haul Trucks, Rail, Transit Buses	Idle Reduction, Retrofits
IN	Northwest Indiana Forum Foundation, Inc.	$164,032		Construction	Replacement/Repower
KS	Kansas Department of Health and Environment	$1,525,524		Delivery Truck, Refuse Hauler, Transit Buses	Retrofits
KY	Kentucky Clean Fuels Coalition	$383,442	$2,160,000	Ports and Airports	Retrofits
MA	Massachusetts Port Authority	$382,397	$106,695	Marine, Ports and Airports	Idle Reduction
MA	Northeast States for Coordinated Air Use Management	$535,250		Rail	Retrofits
MD	Maryland Environmental Services	$178,481		Construction	Retrofits
MD	Montgomery County Public Schools	$699,501	$7,350	School Bus	Retrofits
MI	Lenawee Intermediate School District	$145,337		School Bus	Retrofits

* For FY 2008 grants, the amount listed is the final award amount. For FY 2009 ARRA and FY 2009–2010 grants, the amount listed is the initial award amount.

			FY 2008 Grants		
State	Grant Recipient	EPA Grant Amount	Match	Project Target Fleet(s)	Technology Type(s)
MI	Michigan Clean Energy Coalition	$250,000	$271,690	Construction	Retrofits
MI	NextEnergy Center	$250,000	$152,319	Long Haul Trucks	Idle Reduction
MN	Minnesota Environmental Initiative	$400,000		City/County Vehicle, School Bus, Short Haul, Utility Vehicle	Retrofits
MO	Grace Hill Settlement House	$454,050		School Bus	Retrofits
MO	Missouri Department of Natural Resources	$725,972		Construction	Replacement/Repower
NC	Meckenburg County	$750,000	$750,000	Agriculture, Construction, Long Haul Trucks, Ports and Airports	Retrofits
NC	North Carolina Department of Environment and Natural Resources	$750,000		Construction	Retrofits
NH	Manchester Transit Authority	$229,703	$5,652	Construction, Long Haul Trucks, School Bus, Utility Vehicle	Idle Reduction, Replacement/ Repower, Retrofits
NJ	New Jersey Motor Truck Association	$491,868		Long Haul Trucks	Idle Reduction, Retrofits
NY	Capital District Transportation Authority	$125,000		Transit Buses	Retrofits
NY	Erie County Department of Environmental Planning	$521,667	$72,278	School Bus	Retrofits
NY	Middle Country Central School District	$359,305	$361,305	School Bus	Clean/Alt Fuels
NY	Ulster Board of Cooperative Educational Services	$95,450		School Bus	Retrofits
NY-NJ	Port Authority of New York and New Jersey	$750,000		Long Haul Trucks	Retrofits
NY-NJ	Port Authority of New York and New Jersey	$280,500	$80,500	City/County Vehicle, Utility Vehicle	Retrofits
OH	Clean Fuels Ohio	$412,554		City/County Vehicle, Long Haul Trucks, Utility Vehicle	Retrofits
OH	Ohio Environmental Council	$238,996	$37,660	Ports and Airports, Refuse Hauler, Utility Vehicle	Retrofits
OH	Stark County Educational Service Center	$465,364		School Bus	Retrofits
OR	City of Portland	$393,814		Construction	Replacement/Repower
PA	Pennsylvania Department of Transportation	$219,434		Rail	Idle Reduction
SC	South Carolina State Ports Authority	$729,824	$963,502	Construction, Delivery Truck	Retrofits
SD	Sioux Falls School District	$300,000	$300,000	School Bus	Idle Reduction, Replacement/ Repower, Retrofits
TN	Knox County Government	$40,251	$23,056	Agriculture, Construction, Delivery Truck	Retrofits
TX	North Central Texas Council of Governments	$750,000	$750,000	Construction	Idle Reduction, Replacement/ Repower, Retrofits
TX	North Central Texas Council of Governments	$750,000	$1,500,000	Long Haul Trucks, Utility Vehicle	Idle Reduction
TX	Texas Commission on Environmental Quality	$500,000		School Bus	Retrofits
UT	Utah Dept of Air Quality	$399,955	$4,000	School Bus	Retrofits
VA	Virginia Port Authority	$860,711		Rail	Replacement/Repower
VT	Chittenden Solid Waste District	$198,926	$616,569	Refuse Hauler	Retrofits
WA	Puget Sound Clean Air Agency	$850,000		Construction, Ports and Airports	Retrofits
WI	Wisconsin Department of Transportation	$747,419		Construction	Retrofits

The following grants were closed before work was performed, per grantee and EPA agreement. These recipients returned the EPA award money, which was then reprogammed to fund DERA 2009/2010 grants.

State	Grant Recipient	EPA Grant Amount	Match	Project Target Fleet(s)	Technology Type(s)
AZ	City of Phoenix			Transit Bus	Replacement/Repower
IL	Chicago Public Schools			School Bus	Retrofits
NY	Scarsdale Union Free School District			School Bus	Retrofits

FY 2009 ARRA Grants					
State	Grant Recipient	EPA Grant Amount	Match	Project Target Fleet(s)	Technology Type(s)
AL, FL, GA, KY, MS, NC, SC, TN	American Lung Association	$1.2 million	$1,232,785	Long Haul Trucks	Idle Reduction, Replacement/ Repower, Retrofits
AR	Arkansas Department of Environmental Quality	$793,566	$100,541	Construction	Idle Reduction, Replacement/ Repower, Retrofits
AZ	City of Phoenix Department of Public Works	$829,697		City/County Vehicle	Idle Reduction, Replacement/ Repower, Retrofits
CA	Bay Area Air Quality Management District (BAAQMD)	$2 million	$4,162,636	Long Haul Trucks, Short Haul	Retrofits
CA	California Air Resources Board	$8.89 million		Rail	Idle Reduction, Replacement/ Repower, Retrofits
CA	California Department of Transportation	$951,431	$52,456	Construction	Retrofits
CA	City of Los Angeles Harbor Department	$1.99 million	$675,250	Delivery Truck, Marine, Ports and Airports	Idle Reduction, Replacement/ Repower, Retrofits
CA	Port of Long Beach	$4.01 million		Ports and Airports	Retrofits
CA	San Diego County Air Pollution Control District	$1.56 million		School Bus	Retrofits
CA	San Joaquin Valley Unified Air Pollution Control District	$2 million		Agriculture	Idle Reduction, Replacement/ Repower, Retrofits
CA	San Joaquin Valley Unified Air Pollution Control District	$4 million	$39,150,000	School Bus	Retrofits
CO	City and County of Denver	$700,000		City/County Vehicle, Construction, Refuse Hauler	Idle Reduction, Replacement/ Repower, Retrofits
CO	Colorado Department of Public Health and Environment	$850,000	$774,000	Long Haul Trucks	Idle Reduction
CO	Regional Air Quality Council	$1.25 million		Construction, Long Haul Trucks, Refuse Hauler, School Bus, Utility Vehicle, Delivery Truck, Ports and Airports, City/County Vehicle	Idle Reduction, Replacement/ Repower, Retrofits
CO, MT, ND, SD, UT, WY	Cascade Sierra Solutions	$850,000		Long Haul Trucks	Retrofits
CT	Northeast States for Coordinated Air Use Management	$1.05 million		Rail	Idle Reduction, Replacement/ Repower, Retrofits
CT, MA, ME, NH, RI, VT	Cascade Sierra Solutions (SmartWay Rebate)	$1.15 million		Long Haul Trucks	Idle Reduction
FL	City of Miami	$731,850	$2,195,550	Refuse Hauler	Idle Reduction, Replacement/ Repower, Retrofits
FL	Leon County School Board	$347,288	$1,408,004	School Bus	Idle Reduction, Replacement/ Repower, Retrofits
FL	Miami-Dade Agriculture	$2 million		Agriculture	Idle Reduction, Replacement/ Repower, Retrofits

FY 2009 ARRA Grants

State	Grant Recipient	EPA Grant Amount	Match	Project Target Fleet(s)	Technology Type(s)
FL	Miami-Dade County	$731,850	$2,243,150	Transit Buses	Idle Reduction, Replacement/Repower, Retrofits
GA	Cobb County Schools	$829,697	$1,814,382	School Bus	Retrofits
GA	Georgia Department of Natural Resources	$748,000	$187,000	Long Haul Trucks	Idle Reduction, Replacement/Repower, Retrofits
GA	Georgia Port Authority	$164,000		Ports and Airports	Retrofits
GA	University of Georgia Research Foundation, Inc.	$1.71 million		City/County Vehicle, Emergency Vehicle, School Bus, Transit Buses, Utility Vehicle	Idle Reduction, Replacement/Repower, Retrofits
IL	City of Chicago Department of the Environment	$1 million	$337,500	City/County Vehicle, Construction, Long Haul Trucks, Short Haul, Utility Vehicle	Idle Reduction, Retrofits
IL	Illinois Environmental Protection Agency	$4.17 million	$2,072,744	Construction, Delivery Truck, Long Haul Trucks, Ports and Airports, Rail, School Bus, Transit Buses	Idle Reduction, Retrofits
IN, MN, OH, WI	American Lung Association—Upper Midwest	$3.72 million	$469,000	City/County Vehicle, Construction, Delivery Truck, Long Haul Trucks, School Bus, Short Haul, Transit Buses, Other	Idle Reduction, Retrofits
KS	Johnson County Kansas Government	$1 million	$2,616,605	City/County Vehicle, Construction, Emergency Vehicle, Refuse Hauler	Idle Reduction, Replacement/Repower, Retrofits
KS	Kansas Department of Health and the Environment	$4 million	$1,842,250	Construction, Long Haul Trucks, Rail, Short Haul	Idle Reduction, Replacement/Repower, Retrofits
KY	Kentucky Association of General Contractors	$2 million		Construction	Idle Reduction, Replacement/Repower, Retrofits
LA	Railroad Research Foundation	$2.93 million	$770,993	Rail	Idle Reduction, Replacement/Repower, Retrofits
MA	Chelsea Collaborative	$1.56 million	$10,000	Construction, Delivery Truck, Long Haul Trucks, Short Haul, Utility Vehicle	Retrofits,
MA	Chelsea Collaborative	$357,946	$484,870	Construction, Ports and Airports, Stationary, Other	Idle Reduction
MA	Massachusetts Department of Environmental Protection	$502,500		Short Haul, Utility Vehicle	Retrofits
MD	Maryland Department of the Environment	$1 million		School Bus	Retrofits
MD	Port of Baltimore (Maryland Environmental Services)	$3.5 million		Long Haul Trucks, Marine, Ports and Airports, Rail	Idle Reduction, Replacement/Repower, Retrofits
MD, PA, VA	Mid-Atlantic Regional Air Management Association	$4.32 million	$4,294,366	Long Haul Trucks, Marine, Ports and Airports, Rail, Transit Buses	Retrofits
MD, VA	Chesapeake Bay Foundation	$1.3 million	$200,000	Marine	Retrofits
ME	Maine Department of Environmental Protection	$746,715	$238,088	Construction, Delivery Truck, Ports and Airports, Short Haul, Utility Vehicle, Other	Retrofits
ME	Maine Turnpike Authority	$1.21 million		Long Haul Trucks	Idle Reduction
ME, NH, VT	Northeast States for Coordinated Air Use Management	$1.65 million		Marine, Ports and Airports	Idle Reduction, Replacement/Repower, Retrofits
MI	Great Lakes Commission	$1.21 million	$403,016	Marine, Ports and Airports	Idle Reduction, Replacement/Repower, Retrofits
MI	Greater Lansing Area Clean Cities	$1.07 million	$403,016	School Bus	Retrofits
MI	Lenawee School District	$1.23 million		School Bus	Retrofits
MN	Minnesota Environmental Initiative	$3 million	$489,185	Long Haul Trucks, Rail, School Bus, Short Haul	Idle Reduction
MO	Grace Hill	$2 million	$376,252	Delivery Truck, Emergency Vehicle, Long Haul Trucks, Marine, Ports and Airports, Refuse Hauler, School Bus	Idle Reduction, Replacement/Repower, Retrofits

FY 2009 ARRA Grants

State	Grant Recipient	EPA Grant Amount	Match	Project Target Fleet(s)	Technology Type(s)
MO	Missouri Department of Natural Resources	$975,609	$2,401,550	Refuse Hauler, School Bus, Short Haul	Idle Reduction, Replacement/ Repower, Retrofits
MS	Columbus Municipal School District	$1.45 million	$1,377,177	School Bus	Idle Reduction, Replacement/ Repower, Retrofits
MT	Montana Department of Environmental Quality and Decker Coal	$700,000	$233,333	Construction, Mining	Idle Reduction, Replacement/ Repower, Retrofits
NC	Mecklenburg County	$1.12 million	$1,116,600	Agriculture, Construction, Delivery Truck, Long Haul Trucks, Ports and Airports, Refuse Hauler	Idle Reduction, Replacement/ Repower, Retrofits
NC	North Carolina Department of Public Instruction	$509,000	$267,000	School Bus	Retrofits
ND	Mid-Dakota Education Cooperative	$450,000	$524,970	School Bus	Idle Reduction
NE	Lincoln-Lancaster County Health Department	$1 million	$123,115	Rail, School Bus, Short Haul	Idle Reduction, Replacement/ Repower, Retrofits
NE	University of Nebraska-Lincoln Nebraska Transportation Center (UNL/NTC)	$1 million	$3,131,800	Long Haul Trucks	Idle Reduction, Replacement/ Repower, Retrofits
NJ	Cascade Sierra Solutions	$1.4 million		Long Haul Trucks	Idle Reduction, Replacement/ Repower, Retrofits
NJ, NY	Port Authority of New York and New Jersey	$7 million	$21,000,000	Ports and Airports, Short Haul	Retrofits
NJ, NY, PR	Northeast States for Coordinated Air Use Management	$2.8 million	$1,035,000	Marine	Retrofits
NY	CALSTART, Inc.	$1.3 million	$3,516,156	Delivery Truck	Retrofits
NY	Columbia University	$2 million		Construction, Long Haul Trucks, Short Haul	Retrofits
NY	New York State Department of Transportation	$1.05 million	$400,000	Rail	Retrofits
NY	Port Authority of New York and New Jersey	$2.86 million	$1,441,800	Marine, Ports and Airports	Idle Reduction
OH	Hamilton County Environmental Services	$1.08 million	$3,240,000	School Bus	Retrofits
OH	Ohio Department of Development	$5 million	$2,765,331	Construction, Long Haul Trucks, Ports and Airports, Rail, School Bus	Retrofits
OK	Oklahoma Department of Environmental Quality	$1.85 million		School Bus	Retrofits
OR	Cascade Sierra Solutions—Trailer Aerodynamics Program	$907,072		Long Haul Trucks	Idle Reduction, Replacement/ Repower, Retrofits
OR	City of Portland	$1.62 million		City/County Vehicle, Construction	Idle Reduction, Retrofits
PA	Allegheny County Health Department	$3.5 million		Construction, Rail, Transit Buses	Retrofits
PA	Department of Environmental Protection	$1.5 million		Rail	Retrofits
RI	City of Providence	$565,100		Construction, Short Haul, Utility Vehicle	Retrofits
SC	South Carolina Department of Education	$553,918	$273,772	School Bus	Idle Reduction
SC	South Carolina Port Authority	$2 million	$685,788	Marine, Ports and Airports, Short Haul	Idle Reduction, Replacement/ Repower, Retrofits
SD	Sioux Falls School District	$500,000	$430,000	Construction, School Bus	Idle Reduction, Replacement/ Repower, Retrofits
TN	East Tennessee Clean Fuel Coalition Crossville	$581,849	$239,890	Long Haul Trucks	Idle Reduction
TN	Tennessee Department of Transportation	$2 million		Long Haul Trucks	Idle Reduction, Replacement/ Repower, Retrofits
TX	City of Houston	$2.37 million	$7,097,130	Refuse Hauler	Idle Reduction, Replacement/ Repower, Retrofits
TX	North Central Texas Council of Governments	$2.19 million	$7,187,024	Long Haul Trucks	Idle Reduction, Replacement/ Repower, Retrofits
TX	North Central Texas Council of Governments	$1.55 million		Long Haul Trucks	Idle Reduction, Replacement/ Repower, Retrofits

FY 2009 ARRA Grants

State	Grant Recipient	EPA Grant Amount	Match	Project Target Fleet(s)	Technology Type(s)
TX	Port of Houston Authority	$2.86 million	$3,586,281	Construction, Marine, Ports and Airports	Idle Reduction, Replacement/Repower, Retrofits
TX	Port of Houston Authority	$611,466	$1,377,146	Ports and Airports	Retrofits
TX	VIA Metropolitan Transit	$1.6 million		Transit Buses	Idle Reduction, Replacement/Repower, Retrofits
UT	Utah Department of Environmental Quality	$750,000	$406,000	Agriculture Long Haul Trucks, Delivery Truck	Idle Reduction, Retrofits
VA	Virginia Clean Cities (Hampton Roads)	$1 million	$8,125	Refuse Hauler, School Bus, Transit Buses	Retrofits
WA	Port of Tacoma	$1.49 million	$283,389	Marine, Ports and Airports	Idle Reduction
WA	Puget Sound Clean Air Agency	$2.53 million		Rail	Idle Reduction, Replacement/Repower, Retrofits
WI	Wisconsin Department of Commerce	$2.07 million		Long Haul Trucks	Idle Reduction
WI	Wisconsin Department of Natural Resources	$571,107		Rail	Idle Reduction
WI, MN	National School Transportation Association	$2.42 million	$5,559,958	School Bus	Idle Reduction
WY	Wyoming Department of Environmental Quality	$1.34 million	$777,480	Construction	Retrofits

FY 2009–2010 Grants

State	Grant Recipient	EPA Grant Amount	Match	Project Target Fleet(s)	Technology Type(s)
AR, IA, KS, LA, NE, TX	Owner-Operator Independent Drivers Association	$1 million		Long Haul Trucks	Idle Reduction
AZ	Arizona Department of Commerce	$610,828	$1,347,433	Delivery Truck, Long Haul Trucks	Clean/Alt Fuels, Retrofits
CA	California Air Resources Board	$3.9 million	$1,050,504	Rail	Idle Reduction, Replacement/Repower, Retrofits
CA	City of Long Beach City Harbor Craft and Cargo-Handling	$1.65 million		Marine, Ports and Airports, Utility Vehicle	Idle Reduction, Replacement/Repower, Retrofits
CA	City of Los Angeles Harbor Department	$1.21 million		Marine, Ports and Airports	Idle Reduction, Replacement/Repower, Retrofits
CA	Port of San Francisco	$1 million		Marine, Ports and Airports	Idle Reduction, Replacement/Repower, Retrofits
CA	San Joaquin Valley Unified Air Pollution Control District	$1.80 million		Long Haul Trucks	Idle Reduction, Replacement/Repower, Retrofits
CA	South Coast Air Quality Management District	$1.07 million		School Bus	Idle Reduction, Replacement/Repower, Retrofits
CA (takes place in NJ/NY)	CALSTART, Inc.	$500,576	$432,000	Long Haul Trucks	Idle Reduction, Replacement/Repower, Retrofits
CO	Regional Air Quality Council Comprehensive Clear Skies and Climate Collaborative	$2.50 million	$240,000	City/County Vehicle, Long Haul Trucks, School Bus, Utility Vehicle, Other	Idle Reduction, Replacement/Repower, Retrofits
CT	Fairfield Connecticut Municipal Vehicle Retrofit Project	$831,030		City/County Vehicle, Construction, Refuse Hauler, Utility Vehicle	Idle Reduction, Replacement/Repower, Retrofits
DC	Metro Washington Council of Governments	$560,600	$94,060	Construction, Marine	Retrofits
FL	Florida Refrigeration and Air Conditioning Contractors	$1.68 million	$1,272,250	Construction	Retrofits
FL	Miami-Dade County Miami Port Authority	$1.51 million	$677,796	Ports and Airports	Idle Reduction, Replacement/Repower, Retrofits
GA	University of Georgia Research Foundation, Inc.	$2.72 million	$906,770	Ports and Airports	Idle Reduction, Replacement/Repower, Retrofits

		FY 2009–2010 Grants			
State	Grant Recipient	EPA Grant Amount	Match	Project Target Fleet(s)	Technology Type(s)
HI	Honolulu Clean Cities Hawaii Clean Diesel Initiative	$300,000		Construction, Transit Buses	Idle Reduction, Replacement/ Repower, Retrofits
IL	American Lung Association of the Upper Midwest American Lung Association	$1.5 million	$387,492	Long Haul Trucks, School Bus	Idle Reduction, Replacement/ Repower, Retrofits
IL	Metropolitan Mayors Caucus	$500,000	$3,702,977	School Bus	Idle Reduction, Replacement/ Repower, Retrofits
IN	South Shore Clean Cities, Inc.	$630,500	$427,000	Marine	Idle Reduction, Replacement/ Repower, Retrofits
KS	Kansas Department of Health	$233,218		Construction, Delivery Truck, Long Haul Trucks, School Bus, Short Haul	Idle Reduction, Replacement/ Repower, Retrofits
KY	Louisville/Jefferson County Metro Government	$1.16 million		Construction, Ports and Airports, Transit Buses	Idle Reduction, Replacement/ Repower, Retrofits
LA	Regional Planning Commission for Jefferson, Orleans, Parishes	$1.05 million	$350,000	Rail	Idle Reduction, Replacement/ Repower, Retrofits
MA	Harbor Development Commission Port of New Bedford Shore-side Power Electrification Project	$1 million		Marine, Ports and Airports	Idle Reduction, Replacement/ Repower, Retrofits
MA	Massachusetts Bay Transportation Authority	$700,000	$280,000	Rail	Idle Reduction, Replacement/ Repower, Retrofits
MA	Northeast States for Coordinated Air Use Management	$850,000		Rail	Idle Reduction, Replacement/ Repower, Retrofits
MA (takes place in NJ)	Northeast States for Coordinated Air Use Management	$1.13 million	$333,000	Ports and Airports	Idle Reduction, Replacement/ Repower, Retrofits
MA (takes place in NY/NJ)	Northeast States for Coordinated Air Use Management	$1.42 million	$408,000	Construction, Ports and Airports	Idle Reduction, Replacement/ Repower, Retrofits
MD	Maryland Transit Administration	$975,000	$425,000	Rail	Idle Reduction, Replacement/ Repower, Retrofits
MD	Maryland Transit Administration	$505,000	$226,460	Transit Buses	Idle Reduction, Replacement/ Repower, Retrofits
MD	Mid-Atlantic Regional Air Management Association (MARAMA)	$500,000	$282,500	Marine, Ports and Airports	Idle Reduction, Replacement/ Repower, Retrofits
ME	Maine Department of Environmental Protection	$571,638		Marine, Ports and Airports	Idle Reduction, Replacement/ Repower, Retrofits
MI	Clean Energy Coalition Clean Energy Coalition	$600,000	$133,708	Construction	Idle Reduction, Replacement/ Repower, Retrofits
MI	Michigan Infrastructure and Transportation Association	$966,555	$295,385	Construction	Idle Reduction, Replacement/ Repower, Retrofits
MI	Okemos Public Schools	$877,076		School Bus	Idle Reduction, Replacement/ Repower, Retrofits
MN	Minnesota Environmental Initiative	$977,242	$88,875	Construction, Long Haul Trucks, Refuse Hauler, School Bus, Short Haul, Other	Idle Reduction, Replacement/ Repower, Retrofits
MO	Missouri Department of Natural Resources	$1.8 million		Construction, Refuse Hauler, School Bus	Idle Reduction, Retrofits
MO	National School Transportation Association	$362,642	$243,143	School Bus	Idle Reduction
MO	Passenger Vessel Association	$814,491	$400,566	Marine	Idle Reduction, Retrofits
MS	Mississippi State University	$1.1 million		Long Haul Trucks	Idle Reduction, Replacement/ Repower, Retrofits
MT	Missoula County	$1.13 million		Ports and Airports, Rail	Idle Reduction, Replacement/ Repower, Retrofits

\multicolumn{6}{c}{FY 2009–2010 Grants}					
State	Grant Recipient	EPA Grant Amount	Match	Project Target Fleet(s)	Technology Type(s)
NY	Nassau County Police Department	$708,397	$236,132	Marine, Ports and Airports	Idle Reduction, Replacement/Repower, Retrofits
NY	New York City Department of Transportation	$2 million	$730,500	Marine	Idle Reduction, Replacement/Repower, Retrofits
NY	The Port Authority of NY & NJ	$1.58 million	$4,731,447	Ports and Airports	Idle Reduction, Replacement/Repower, Retrofits
OH	Clean Fuels Ohio	$750,000	$345,485	Construction, Delivery Truck, Long Haul Trucks, School Bus, Other	Idle Reduction, Replacement/Repower, Retrofits
OH	Ohio Regional Planning Commission	$1.23 million	$198,850	Construction, Emergency Vehicle, Long Haul Trucks, School Bus, Short Haul, Other	Idle Reduction, Replacement/Repower, Retrofits
OR	Oregon Department of Environmental Quality	$482,476	$461,984	Marine, Ports and Airports	Idle Reduction, Replacement/Repower, Retrofits
PA	City of Philadelphia	$475,669	$4,320,000	Ports and Airports, Transit Buses, Other	Retrofits
PA	Clean Air Council Clean Air Council	$350,000		Ports and Airports, Other	Retrofits
PA	Port of Pittsburgh Commission	$1.16 million		Marine, Ports and Airports	Idle Reduction, Replacement/Repower, Retrofits
PR	Autoridad de Transporte Maritimo	$517,220	$339,522	Marine	Idle Reduction, Replacement/Repower, Retrofits
TX	Cascade Sierra Solutions	$1.15 million		Delivery Truck, Long Haul Trucks	Idle Reduction
TX	Houston Advanced Research Center	$496,000		Long Haul Trucks	Idle Reduction
TX	North Central Texas Council of Governments	$500,000		Construction	Idle Reduction, Replacement/Repower, Retrofits
TX	North Central Texas Council of Governments	$500,000		Long Haul Trucks	Idle Reduction, Replacement/Repower, Retrofits
TX	North Central Texas Council of Governments	$500,000	$159,775	School Bus	Idle Reduction, Replacement/Repower, Retrofits
TX	Port of Corpus Christi Locomotive Switch Engine Repower Project	$1.03 million	$342,019	Marine, Rail	Idle Reduction, Replacement/Repower, Retrofits
TX	Port of Houston Authority	$1.49 million		Marine, Ports and Airports	Idle Reduction, Replacement/Repower, Retrofits
VA	James Madison University Virginia	$710,000	$1,107,503	Construction	Retrofits
VA	National School Transportation Association	$362,642		School Bus	Idle Reduction, Replacement/Repower
VA	Virginia Port Authority Dredging Repower Project	$775,000	$932,750	Marine	Idle Reduction, Replacement/Repower, Retrofits
WA	Pacific Northwest Poll Prev Res Center	$875,972		Long Haul Trucks	Idle Reduction, Replacement/Repower, Retrofits
WA	Washington Department of Ecology	$875,972		School Bus	Idle Reduction
WI	Associated General Contractors of Greater Milwaukee	$500,000	$169,576	City/County Vehicle, Construction, Marine	Idle Reduction, Replacement/Repower, Retrofits
WI	Leonardo Academy, Inc.	$600,000	$353,600	Long Haul Trucks, School Bus, Transit Buses	Idle Reduction, Replacement/Repower, Retrofits
WI	Wisconsin Department of Natural Resources	$1.18 million	$1,816,835	City/County Vehicle, Construction, Long Haul Trucks, Ports and Airports, School Bus, Other	Idle Reduction, Replacement/Repower, Retrofits
WV	Railroad Research Foundation of West Virginia	$975,000	$464,000	Rail	Idle Reduction, Replacement/Repower, Retrofits

FY 2009–2010 Grants, Tribal					
State	Grant Recipient	EPA Grant Amount	Match	Project Target Fleet(s)	Technology Type(s)
CA	Morongo Band of Mission Indians	$250,000	$191,700	City/County Vehicle, Construction, School Bus, Transit Buses	Retrofits
CA	Soboba Band of Luiseno Indians	$78,000	$18,300	School Bus	Retrofits
IA	Meskwaki Nation	$190,000		School Bus	Idle Reduction, Replacement/ Repower, Retrofits
MN	Leech Lake Band of Ojibwa	$134,404	$45,100	Refuse Hauler	Idle Reduction, Replacement/ Repower, Retrofits

Appendix B: Emerging Technologies

FY 2008 Grants			
State	Grant Recipient	EPA Funding	Project Target Fleet(s)
CA	South Coast Air Quality Management District	$900,000	Delivery Truck, Long Haul Trucks
TX	Center for Transport and the Environment	$300,000	Delivery Truck, Short Haul
TX	Texas Transportation Institute	$500,000	Construction
TX	University of Houston	$500,000	Construction
WA	Puget Sound Clean Air Agency	$700,000	Marine

FY 2009 ARRA Grants			
State	Grant Recipient	EPA Funding	Project Target Fleet(s)
CA	San Joaquin Valley Air Pollution Control District	$1.3 million	Long Haul Trucks
CA	South Coast Air Quality Management District	$2 million	Long Haul Trucks
CA	South Coast Air Quality Management District	$2 million	Long Haul Trucks
IN	Indiana Department of Environmental Management	$1 million	Construction
NC	Southern Research Institute	$1 million	Refuse Hauler
PA	Pennsylvania Department of Environmental Protection	$1.5 million	Marine, Ports and Airports
TN	Mississippi River Corridor—Tennessee, Inc.	$2 million	Marine, Ports and Airports
TX	City of Irving	$937,605	Refuse Hauler, Utility Vehicle
TX	Houston Advanced Research Center	$2.36 million	Construction, Delivery Truck
TX	Houston Advanced Research Center	$1.6 million	Marine, Ports and Airports
TX	University of Houston	$1.74 million	City/County Vehicle, School Bus, Utility Vehicle
TX	University of Houston	$1.8 million	School Bus
VA	Fairfax County	$1.3 million	School Bus
WA	Puget Sound Clean Air Agency	$42,000	Marine, Ports and Airports

FY 2009–2010 Grants			
State	Grant Recipient	EPA Funding	Project Target Fleet(s)
CA	City of Los Angeles Harbor Department	$731,298	Ports and Airports
CA	California Air Resources Board	$1.2 million	Rail
CA	South Coast Air Quality Management District	$1.5 million	Marine, Ports and Airports
TX	University of Houston	$1.05 million	School Bus
WA	Puget Sound Clean Air Agency	$1.2 million	Marine, Ports and Airports

Appendix C: SmartWay Finance

FY 2008 Grants				
State	Grant Recipient	EPA Funding	Project Target Fleet(s)	Technology Type(s)
CA, NY, OR, WA	Cascade Sierra Solutions	$1.13 million	Long Haul Trucks, Ports and Airports	Idle Reduction, Retrofits
Nationwide	Community Development Transportation Lending Services	$1.13 million	Long Haul Trucks, Ports and Airports	Idle Reduction, Retrofits
Nationwide	Owner-Operator Independent Drivers Association	$1.13 million	Long Haul Trucks	Idle Reduction

FY 2009 ARRA Grants				
State	Grant Recipient	EPA Funding	Project Target Fleet(s)	Technology Type(s)
CA	California Air Resources Board	$5 million	Construction, Marine	Idle Reduction, Replacements, Retrofits
CA, OR, WA	Cascade Sierra Solutions	$9 million	Long Haul Trucks, Ports and Airports	Replacements, Retrofits
KY	Louisville-Jefferson County Metro Government	$2 million	Construction	Idle Reduction, Replacements, Retrofits
TX	Houston-Galveston Area Council	$9 million	Ports and Airports	Replacements, Retrofits
Nationwide	National Association for Pupil Transportation	$5 million	School Bus	Idle Reduction, Replacements, Retrofits

FY 2009–2010 Grants				
State	Grant Recipient	EPA Funding	Project Target Fleet(s)	Technology Type(s)
MD, DE, PA, VA	Mid-Atlantic Regional Air Management Association	$3.92 million	Ports and Airports	Replacements
TN, GA	Southern Alliance for Clean Energy	$5 million	Long Haul Trucks	Idle Reduction, Retrofits
Nationwide	Cascade Sierra Solutions	$2 million	Long Haul Trucks, Ports and Airports	Replacements
Nationwide	Community Development Transportation Lending Services	$2.6 million	Long Haul Trucks, Ports and Airports	Idle Reduction, Retrofits

Appendix D: State Clean Diesel Programs

State	Grant Recipient	FY 2008 Total Federal Funding Allotment	FY 2009 Total Federal Funding Allotment	FY 2010 Total Federal Funding Allotment	Recovery Act Total Federal Funding Allotment	Sector(s)	Technology Types
AK	Alaska Department of Environmental Conservation	$196,880	$235,294	$352,941	$1,730,000	Stationary, City/county vehicles, Long-haul trucks, Locomotives, School bus	Idle reduction, Replacement/repower, Retrofit
AL	Alabama Department of Environmental Management	$295,320	$235,294	$235,294	$1,730,000	Construction, Locomotives, City/county vehicles, Utility vehicles	Diesel particulate filters, Clean fuels, Idle reduction
AR	Arkansas Department of Environmental Quality	$295,320	$352,941	$235,294	$1,730,000	School buses, Construction, Long-haul trucks	Diesel oxidation catalysts, Auxiliary power units, Replacement/repower
AZ	Arizona Department of Environmental Quality	$196,880	$235,294	$235,200	$1,730,000	School buses, Long-haul trucks	Retrofit, Truck stop electrification
CA	California Air Resources Board	$295,320	$352,941	$352,941	$1,730,000	City/county vehicles, Utility vehicles, Refuse haulers	Idle reduction, Diesel particulate filters, Replacement/repower
CO	Colorado Department of Public Health and Environment	$196,880	$235,294	$235,294	$1,730,000	School buses	Idle reduction, Replacement/repower, Diesel oxidation catalysts, Closed crankcase filtration units
CT	Connecticut Department of Environmental Protection	$295,320	$235,294	$235,294	$1,730,000	City/County vehicles, Delivery trucks, Long-haul trucks, Marine, Ports and airports, School buses, Short-haul trucks, Transit buses, Utility vehicles, Construction, Locomotives	Replacement/repower, Diesel oxidation catalysts, Truck stop electrification
DC	District Department of the Environment		$235,294	$235,294	$1,730,000	Transit buses, Utility vehicles, Refuse haulers, City/county vehicles, Marine	Replacement/repower, Clean/alternate fuel
DE	Delaware Department of Natural Resources and Environmental Control	$295,320	$352,941	$352,941	$1,730,000	Agriculture, Construction, School buses, Utility vehicles, Long-haul trucks, Delivery trucks, Ports and airports, Stationary, City/county vehicles, Refuse haulers	Diesel particulate filters, Diesel oxidation catalysts, Auxiliary power units, Replacement/repower
FL	Florida Department of Environmental Control	$295,320	$352,941	$352,941	$1,730,000	Long-haul trucks, School buses	Diesel oxidation catalysts, Auxiliary power units, Truck stop electrification
GA	Georgia Department of Natural Resources	$295,320	$352,941	$352,941	$1,730,000	School buses	Closed crankcase, Partial flow filters, Clean/alternative fuels,
HI	Hawaii Department of Health	$196,880	$235,294	$235,294	$1,730,000	School buses, Refuse haulers, City/county vehicles, Utility vehicles	Replacement/repower, Idle reduction, Diesel particulate filters
IA	Iowa Department of Natural Resources	$196,880	$235,294	$352,941	$1,730,000	School buses, Construction, Long-haul trucks, Locomotives, Transit buses	Idle reduction, Retrofits
ID	Idaho Department of Environmental Quality	$196,880	$235,294	$235,294	$1,730,000	School buses	Diesel oxidation catalysts, Closed crankcase ventilation, Direct fired heaters, Replacement/repower
IL	Illinois Environmental Protection Agency	$295,320	$352,941	$352,941	$1,730,000	Construction, Transit buses, School buses, Long-haul trucks, Locomotives, Refuse haulers	Diesel oxidation catalysts, Diesel particulate filters, Auxiliary power units, Replacement/repower

State	Grant Recipient	FY 2008 Total Federal Funding Allotment	FY 2009 Total Federal Funding Allotment	FY 2010 Total Federal Funding Allotment	Recovery Act Total Federal Funding Allotment	Sector(s)	Technology Types
IN	Indiana Department of Environmental Management	$196,880	$235,294	$352,941	$1,730,000	City/county vehicles, Construction, Long-haul trucks, Locomotives, Refuse haulers, School buses, Transit buses, Utility vehicles	Diesel oxidation catalysts, Diesel particulate filters, Auxiliary power units, Replacement/repower
KS	Kansas Department of Health and Environment	$196,880	$352,941	$352,941	$1,730,000	School buses, Long-haul trucks, Agriculture, City/county vehicles, Construction, Delivery trucks	Auxiliary power units, Replacement/repower, Diesel oxidation catalysts, Closed crankcase ventilation, Partial flow filters, Diesel particulate filters
KY	Kentucky Division for Air Quality	$196,880	$235,294	$235,294	$1,730,000	School buses, Long-haul trucks, Transit buses	Diesel oxidation catalysts, Diesel particulate filters, Closed crankcase ventilation, Auxiliary power units
LA	Louisiana Department of Environmental Quality	$295,320	$352,941	$235,294	$1,730,000	Marine, Agriculture, Delivery trucks	Idle reduction, Replacement/repower, Retrofits
MA	Massachusetts Department of Environmental Protection	$295,320	$352,941	$352,941	$1,730,000	City/County vehicles, Delivery trucks, Locomotives, Marine, Refuse haulers, School buses, Short-haul, Transit buses, Utility vehicles, Long-haul trucks	Diesel oxidation catalysts, Replacement/repower
MD	Maryland Department of the Environment	$295,320	$235,294	$235,294	$1,730,000	Marine, School buses, Transit buses, Refuse haulers	Diesel particulate filters, Closed crankcase ventilation, Idle reduction, Replacement/repower
ME	Maine Department of Environmental Protection	$295,320	$352,941	$352,941	$1,730,000	City/county vehicles, Long-haul trucks, Marine, Locomotives, School buses, Utility vehicles	Idle reduction, Diesel oxidation catalysts, Replacement/repower, Clean/alternate fuels
MI	Michigan Department of Environmental Quality	$295,320	$352,941	$352,941	$1,730,000	Refuse haulers, School buses	Diesel oxidation catalysts
MN	Minnesota Pollution Control Agency	$295,320	$352,941	$352,941	$1,730,000	Long-haul trucks, School buses, Construction, Transit buses	Replacement/repower, Auxiliary power units, Diesel oxidation catalysts, Diesel particulate filters
MO	Missouri Department of Natural Resources	$295,320	$352,941	$352,941	$1,730,000	Construction, Delivery trucks, Marine, Locomotives, Refuse haulers, School buses, Short-haul trucks, Long-haul trucks	Idle reduction, Replacement/repower, Retrofits
MS	Mississippi Department of Environmental Quality	$295,320	$352,941	$352,941	$1,730,000	School buses, City/county vehicles, Utility vehicles	Diesel oxidation catalysts
MT	Montana Department of Environmental Quality	$295,320	$352,941	$352,941	$1,730,000	School buses	Replacement/repower, Idle reduction, Retrofits
NC	North Carolina Department of the Environment and Natural Resources	$295,320	$352,941	$352,941	$1,730,000	Port and airports, School buses, Marine, Construction, Locomotives, Refuse haulers, Transit buses, Long-haul trucks, Emergency vehicles	Replacement/repower, Idle reduction, Retrofits
ND	North Dakota Department of Health	$196,880	$352,941	$352,941	$1,730,000	School buses, City/county vehicles	Replacement/repower, Idle reduction, Retrofits
NE	Nebraska Department of Environmental Quality	$196,880	$235,294	$235,294	$1,730,000	School buses, City/county vehicles, Construction, Refuse haulers, Transit buses, Utility vehicles, Short-haul trucks, Long-haul trucks, Emergency vehicles	Idle reduction, Replacement/repower, Retrofits
NH	New Hampshire Department of Environmental Science	$295,320	$235,294	$235,294	$1,730,000	Locomotives, Marine, School buses, Transit buses, City/county vehicles, Construction, Delivery trucks, Long-haul trucks, Ports and airports, Refuse haulers	Auxiliary power units, Diesel particulate filters, Replacement/repower
NJ	New Jersey Department of Environmental Protection	$295,320	$352,941	$352,941	$1,730,000	Construction, Delivery trucks, Ports and airports	Diesel particulate filters, Diesel oxidation catalysts
NM	New Mexico Environment Department	$196,880	$235,294	$235,294	$1,730,000	Transit buses, Construction, School buses	Auxiliary power units, Diesel particulate filters, Replacement/repower

State	Grant Recipient	FY 2008 Total Federal Funding Allotment	FY 2009 Total Federal Funding Allotment	FY 2010 Total Federal Funding Allotment	Recovery Act Total Federal Funding Allotment	Sector(s)	Technology Types
NV	Nevada Division of Environmental Protection	$295,320	$352,941	$352,941	$1,730,000	School buses	Replacement/repower, Idle reduction, Retrofits
NY	New York State Department of Environmental Conservation	$295,320	$352,941	$352,941	$1,730,000	Transit buses, School buses	Diesel oxidation catalysts, Closed crankcase ventilation
OH	Ohio Environment Protection Agency	$295,320	$352,941	$352,941	$1,730,000	School buses	Diesel oxidation catalysts, Closed crankcase ventilation, Diesel particulate filters
OK	Oklahoma Department of Environmental Quality	$295,320	$235,294	$235,200	$1,730,000	School buses	Retrofits
OR	Oregon Department of Environmental Quality	$295,320	$352,941	$352,941	$1,730,000	Construction, Refuse haulers, Marine	Diesel oxidation catalysts, Diesel particulate filters, Replacement/repower
PA	Pennsylvania Department of Environmental Protection	$295,320	$235,294	$352,941	$1,730,000	Construction, Long-haul trucks, Ports and airports, Refuse hauler, School buses, Transit buses	Diesel particulate filters, Replacement/repower, Diesel oxidation catalysts, Closed crankcase ventilation, Clean/alternate fuel
RI	Rhode Island Department of Environmental Management	$196,880	$352,941	$235,294	$1,730,000	City/county vehicles, Marine, Ports and airports, Utility vehicles, School buses, Construction, Refuse haulers	Diesel oxidation catalysts
SC	South Carolina Department of Health and Environmental Control	$295,320	$352,941	$352,941	$1,730,000	Construction, Refuse haulers, City/county vehicles, School buses	Replacement/repower, Diesel oxidation catalysts, Closed crankcase ventilation, Diesel particulate filters
SD	South Dakota Department of Environment and Natural Resources	$196,880	$235,294	$235,294	$1,730,000	School buses	Replacement/repower, Diesel oxidation catalysts, Idle reduction
TN	Tennessee Department of Environment and Conservation	$295,320	$235,294	$235,294	$1,730,000	Long-haul trucks	Auxiliary power units
TX	Texas Commission on Environmental Quality	$295,320	$235,294	$235,294	$1,730,000	School buses	Idle reduction, Replacement/repower, Retrofits
UT	Utah Department of Environmental Quality	$295,320	$352,941	$352,941	$1,730,000	Delivery trucks, School buses	Clean/alternate fuel, Replacement/repower, Diesel oxidation catalysts, Closed crankcase ventilation
VA	Virginia Department of Environmental Quality	$196,880	$235,294	$235,294	$1,730,000	City/county vehicles, Refuse haulers, School buses, Marine, Short-haul trucks	Idle reduction, Replacement/repower, Diesel oxidation catalysts
VT	Vermont Department of Environmental Conservation	$196,880	$235,294	$235,294	$1,730,000	School buses, City/county vehicles, Construction	Replacement/repower, Idle reduction, Retrofits
WA	Washington State Department of Ecology	$295,320	$352,941	$0	$1,730,000	Marine, Ports and airports, School buses, Transit buses, Construction	Replacement/repower, Diesel oxidation catalysts, Diesel particulate filters, Idle reduction
WI	Wisconsin Department of Natural Resources	$295,320	$352,941	$352,941	$1,730,000	School buses, Construction	Diesel oxidation catalysts, Replacement/repower
WV	West Virginia Department of Environmental Protection	$196,880	$235,294	$235,294	$1,730,000	School buses, Transit buses, Refuse haulers, Construction	Replacement/repower, Idle reduction, Retrofits, Clean/alternate fuels
WY	Wyoming Department of Environmental Quality	$196,880	$235,294	$352,941	$1,730,000	Construction, School buses	Replacement/repower, Diesel oxidation catalysts, Closed crankcase ventilation, Idle reduction

Appendix E: National Program Evaluation Criteria

For the National Clean Diesel Funding Assistance Program applicants, EPA applied criteria and points such as those summarized in the list below, which are consistent with the priorities described in Section 792 of the Energy Policy Act, Subtitle G, and with Agency policy. Evaluation criteria differed slightly between the National, Emerging Technologies, and SmartWay programs.

- ⊃ Project summary/approach
- ⊃ Programmatic priorities
- ⊃ Past performance
- ⊃ Environmental results
- ⊃ Budget/resources
- ⊃ Target fleet
- ⊃ Leveraging resources and partners
- ⊃ Staff expertise/qualifications
- ⊃ Regional significance

For the Recovery Act grant competition, EPA used the same criteria, but also took job creation/retention and "shovel-ready" projects into consideration.

For more detailed information about the Request for Proposals, please visit www.epa.gov/cleandiesel/prgnational.htm.

Appendix F: Clean Diesel Collaboratives

Clean Diesel Collaboratives are public-private partnerships that include EPA regional offices as well as equipment manufacturers, fleet owners, state and local governments, and nonprofit organizations. They are diverse, multi-stakeholder groups that provide technical assistance, foster partnerships, and identify and leverage resources.

The Northeast Clean Diesel Collaborative: Connecticut, Maine, Massachusetts, New Hampshire, New Jersey, New York, Rhode Island, Vermont, Puerto Rico, and the U.S. Virgin Islands
www.northeastdiesel.org/

The Mid-Atlantic Diesel Collaborative: Delaware, Maryland, Pennsylvania, Virginia, West Virginia, and Washington D.C.
www.dieselmidatlantic.org/

The Southeast Diesel Collaborative: Alabama, Florida, Georgia, Kentucky, Mississippi, North Carolina, South Carolina, and Tennessee
www.southeastdiesel.org/

The Midwest Clean Diesel Initiative: Illinois, Indiana, Michigan, Minnesota, Ohio, and Wisconsin
www.epa.gov/midwestcleandiesel/

The Blue Skyways Collaborative: Arkansas, Iowa, Kansas, Louisiana, Minnesota, Missouri, Nebraska, New Mexico, Oklahoma, and Texas
www.blueskyways.org/

The Rocky Mountain Clean Diesel Collaborative: Colorado, Montana, North Dakota, South Dakota, Utah, and Wyoming
www.epa.gov/region8/air/rmcdc/

The West Coast Diesel Collaborative: Alaska, Arizona, California, Hawaii, Idaho, Nevada, Oregon, Washington, the territories of Guam and American Samoa, and the Commonwealth of the Northern Mariana Islands
www.westcoastcollaborative.org/

Appendix G: Acronyms and Abbreviations

APU	Auxiliary Power Unit	**FY**	Fiscal Year
CARB	California Air Resources Board	**HC**	Hydrocarbon
CCV	Closed Crankcase Ventilation	**NAAQS**	National Ambient Air Quality Standards
CNG	Compressed Natural Gas	**NCDC**	National Clean Diesel Campaign
CO	Carbon Monoxide	**NESCAUM**	Northeast States for Coordinated Air Use Management
CO$_2$	Carbon Dioxide		
DEQ	Diesel Emissions Quantifier	**NO$_x$**	Nitrogen Oxides
DERA	Diesel Emissions Reduction Act	**OIG**	EPA's Office of the Inspector General
DFH	Direct Fired Heater	**PFF**	Partial Flow Filter
DOC	Diesel Oxidation Catalyst	**PM**	Particulate Matter
DPF	Diesel Particulate Filter	**Recovery Act**	American Reinvestment and Recovery Act
DRIVER	Database for Reporting Innovative Emissions Reductions	**RFP**	Request for Proposals
EGR	Exhaust Gas Recirculation	**SCR**	Selective Catalytic Reduction
EPA	U.S. Environmental Protection Agency	**SO$_x$**	Sulfur Oxides
		TSE	Truck Stop Electrification
EPAct	Energy Policy Act	**ULSD**	Ultra Low Sulfur Diesel
ET	Emerging Technologies		